Lecture Notes in Business Information Processing

548

Series Editors

Wil van der Aalst , *RWTH Aachen University, Aachen, Germany*
Sudha Ram , *University of Arizona, Tucson, USA*
Michael Rosemann , *Queensland University of Technology, Brisbane, Australia*
Clemens Szyperski, *Microsoft Research, Redmond, USA*
Giancarlo Guizzardi , *University of Twente, Enschede, The Netherlands*

LNBIP reports state-of-the-art results in areas related to business information systems and industrial application software development – timely, at a high level, and in both printed and electronic form.

The type of material published includes

- Proceedings (published in time for the respective event)
- Postproceedings (consisting of thoroughly revised and/or extended final papers)
- Other edited monographs (such as, for example, project reports or invited volumes)
- Tutorials (coherently integrated collections of lectures given at advanced courses, seminars, schools, etc.)
- Award-winning or exceptional theses

LNBIP is abstracted/indexed in DBLP, EI and Scopus. LNBIP volumes are also submitted for the inclusion in ISI Proceedings.

Jānis Grabis · Tanja E. J. Vos ·
Maria José Escalona · Oscar Pastor
Editors

Research Challenges in Information Science

19th International Conference, RCIS 2025
Seville, Spain, May 20–23, 2025
Proceedings, Part II

 Springer

Editors
Jānis Grabis ⓘ
Riga Technical University
Riga, Latvia

Tanja E. J. Vos ⓘ
Open Universiteit
Heerlen, The Netherlands

Maria José Escalona ⓘ
Universidad de Sevilla
Seville, Spain

Oscar Pastor ⓘ
Universidad Politécnica de Valencia
Valencia, Valencia, Spain

ISSN 1865-1348   ISSN 1865-1356 (electronic)
Lecture Notes in Business Information Processing
ISBN 978-3-031-92470-5   ISBN 978-3-031-92471-2 (eBook)
https://doi.org/10.1007/978-3-031-92471-2

© The Editor(s) (if applicable) and The Author(s), under exclusive license
to Springer Nature Switzerland AG 2025

This work is subject to copyright. All rights are solely and exclusively licensed by the Publisher, whether the whole or part of the material is concerned, specifically the rights of translation, reprinting, reuse of illustrations, recitation, broadcasting, reproduction on microfilms or in any other physical way, and transmission or information storage and retrieval, electronic adaptation, computer software, or by similar or dissimilar methodology now known or hereafter developed.
The use of general descriptive names, registered names, trademarks, service marks, etc. in this publication does not imply, even in the absence of a specific statement, that such names are exempt from the relevant protective laws and regulations and therefore free for general use.
The publisher, the authors and the editors are safe to assume that the advice and information in this book are believed to be true and accurate at the date of publication. Neither the publisher nor the authors or the editors give a warranty, expressed or implied, with respect to the material contained herein or for any errors or omissions that may have been made. The publisher remains neutral with regard to jurisdictional claims in published maps and institutional affiliations.

This Springer imprint is published by the registered company Springer Nature Switzerland AG
The registered company address is: Gewerbestrasse 11, 6330 Cham, Switzerland

If disposing of this product, please recycle the paper.

Preface

Volumes 547 and 548 of the Lecture Notes in Business Information Processing series contain the proceedings of the 19th International Conference on Research Challenges in Information Science, RCIS 2025, held in Seville, the capital of Andalusia in Spain, during 20-23 May, 2025. Seville is famous for its stunning Moorish and Gothic landmarks, including the Alcázar of Seville, a UNESCO World Heritage Site. Also, the city is the birthplace of the flamenco dance and it hosts the world-famous *Feria de Abril*, a lively festival filled with colourful dresses, horse parades, and Andalusian culture.

The scope of RCIS covers a broad range of thematic areas, including information systems and their engineering, user-oriented approaches, data and information management, enterprise management and engineering, domain-specific information systems engineering, data science, information infrastructures, and reflective research and practice. RCIS 2025 focused on the special theme "Advancing Information Science and Information Systems Quality in the Era of Complexity."

We received a total of 103 full paper submissions that were carefully reviewed and selected. Among these, five submissions were desk rejected for being outside the scope of the conference. The remaining papers underwent a rigorous single-blind review process, each evaluated by at least three Program Committee (PC) members. Following this, the review process included a discussion period moderated by Program Board (PB) members, culminating in a PB meeting held in Paris to finalize the selection of papers for the conference program. Eventually 33 papers were accepted, meaning the acceptance rate for the main conference was 34%. Authors were provided with meta-reviews and recommendations to incorporate in their camera-ready versions.

The proceedings also include 10 short papers accepted for presentation at the Forum, which aims to showcase new research ideas and prototypes in information science, and three short papers accepted for discussion at the Doctoral Consortium.

The conference program commenced with workshops, followed by the main conference featuring keynotes, research paper sessions, tutorials, the Forum, the Doctoral Consortium, research projects, and journal-first presentations. RCIS 2025 hosted three distinguished keynote presentations:

Tijs van der Storm, from the CWI, The Netherlands about *The next 700 Low-Code Platforms*.

Rebeca Gutierrez from the University of Seville, Spain, about *Challenges in Forming European Consortia for Ethical and Human-Centered AI: Overcoming Barriers to Collaborative Innovation*.

Giancarlo Guizzardi from the University of Twente, The Netherlands on *Explanation, Semantics, Ontology & Trustworthiness*.

Additionally, three accepted tutorials provided deep dives into timely topics central to the RCIS community.

Tutorial 1. *Problematising and Ideating for Design Science Research*, delivered by John R. Venable, Curtin University, Australia.

Tutorial 2. *Taking control of our privacy when using mobile apps: checks and tools*, delivered by M. Mercedes Martínez-González, Alejandro Pérez-Fuente, Amador Aparicio, and Pablo-Abel Criado-Lozano, Privacy Engineering Research Group, Universidad de Valladolid, Spain.

Tutorial 3. *Introduction to Fractal Enterprise Model (FEM) and FEM toolkit*, delivered by Ilia Bider, Stockholm University, Sweden

We extend our sincere gratitude to all authors who submitted their work to RCIS 2025, as well as to the Program Committee and Program Board members for their dedication in reviewing and discussing the submitted papers. We also wish to thank all members of the Organization Committee and the student volunteers for their invaluable contributions to making this conference a success.

Jānis Grabis
Tanja E. J. Vos

Organization

Conference Chairs

General Chairs

Oscar Pastor	Universitat Politècnica de València, Spain
María José Escalona	Universidad de Sevilla, Spain

Program Chairs

Jānis Grabis	Riga Technical University, Latvia
Tanja E.J. Vos	Open Universiteit, The Netherlands and Universitat Politècnica de València, Spain

Steering Committee

Saïd Assar	Institut Mines-Télécom Business School, France
Marko Bajec	University of Ljubljana, Slovenia
Xavier Franch	Universitat Politècnica de Catalunya, Spain
Renata Guizzardi	University of Twente, The Netherlands
Haralambos Mouratidis	University of Essex, UK
Pericles Loucopoulos	Institute of Digital Innovation and Research, Ireland
Selmin Nurcan	Université Paris 1 Panthéon-Sorbonne, France
Oscar Pastor	Universitat Politècnica de València, Spain
Jolita Ralyté	University of Geneva, Switzerland
Colette Rolland	Paris1 Panthéon Sorbonne, France
Maribel Yasmina Santos	University of Minho, Portugal
Jelena Zdravkovic	Stockholm University, Sweden

Local Organizing Team

Lola de Acuña	University of Seville, Spain
José Luis Alonso-Rocha	University of Seville, Spain
Irene Barba	University of Seville, Spain
Elena Enamorado	University of Seville, Spain
Julián García-García	University of Seville, Spain
Manuel García Romero	University of Seville, Spain
José Gonzalez-Enríquez	University of Seville, Spain
Javier Gutiérrez	University of Seville, Spain
Andrés Jimenez	University of Seville, Spain
Antonio Martinez-Rojas	University of Seville, Spain
Ana Jaleh Mehrabi de Acuña	University of Seville, Spain
Leticia Morales	University of Seville, Spain
Antonio Rodríguez-Ruiz	University of Seville, Spain
Nicolás Sánchez	University of Seville, Spain
Tomasz Wojdynski	University of Seville, Spain

Doctoral Consortium Chairs

João Araújo	Universidade NOVA de Lisboa, Portugal
Nora Koch	Universidad de Sevilla, Spain

Forum Chairs

Greta Adamo	Universitat Politècnica de València, Spain
Beatriz Marín	Universitat Politècnica de València, Spain

Workshop Chairs

Raian Ali	Hamad Bin Khalifa University, Qatar
Sergio España	Utrecht University, The Netherlands
Sameha Alshakhsi	Hamad Bin Khalifa University, Qatar

Tutorial Chairs

Rebecca Deneckere   Centre de Recherche en Informatique, France
Mohamad Gharib   University of Tartu, Estonia

Research Projects Chairs

Isabel Brito   Polytechnic Institute of Beja, Portugal
Irina Rychkova   Université Paris 1 Panthéon-Sorbonne, France

Journal-First Chairs

Saïd Assar   Institut Mines-Télécom Business School, France
Iris Reinhartz-Berger   University of Haifa, Israel

Publicity Chairs

Istvan David   McMaster University, Canada
Ben Roelens   Open Universiteit, The Netherlands
Pedro Guimaraes   Universidade do Minho, Portugal
Glenda Amaral   University of Twente, The Netherlands

Student Internationalization Chair

Selmin Nurcan   Université Paris 1 Panthéon - Sorbonne, France

Proceedings Chairs

Elizabete Citskovska   Riga Technical University, Latvia

Program Board

Saïd Assar   Institut Mines-Télécom Business School, France
Marko Bajec   University of Ljubljana, Slovenia
Dominik Bork   TU Wien, Austria

Xavier Franch — Universitat Politècnica de Catalunya, Spain
Renata Guizzardi — University of Twente, The Netherlands
Haralambos Mouratidis — University of Essex, UK
Pericles Loucopoulos — Institute of Digital Innovation and Research, Ireland
Selmin Nurcan — Université Paris 1 Panthéon-Sorbonne, France
Oscar Pastor — Universitat Politècnica de València, Spain
Jolita Ralyté — University of Geneva, Switzerland
Manfred Reichert — University of Ulm, Germany
Colette Rolland — Paris1 Panthéon Sorbonne, France
Maribel Yasmina Santos — University of Minho, Portugal
Jelena Zdravkovic — Stockholm University, Sweden

Program Committee

Ademar Aguiar — University of Porto, Portugal
Raian Ali — Hamad Bin Khalifa University, Qatar
Carina Alves — Universidade Federal de Pernambuco, Brazil
Vasco Amaral — NOVA University Lisbon, Portugal
João Araújo — Universidade NOVA de Lisboa, Portugal
Fatma Başak Aydemir — Utrecht University, The Netherlands
Clara Ayora — Universidad de Castilla-La Mancha, Spain
Anna Bernasconi — Politecnico di Milano, Italy
Isabel Sofia Brito — Polytechnic Institute of Beja, Portugal
Jean-Michel Bruel — IRIT, France
Faiza Bukhsh — University of Twente, The Netehrlands
Nelly Condori Fernández — Universidad de Santiago de Compostela, Spain
Mario Cortes-Cornax — Université Grenoble Alpes, France
Fabiano Dalpiaz — Utrecht University, The Netherlands
Maya Daneva — University of Twente, The Netherlands
Istvan David — McMaster University, Canada
Andrea Delgado — Universidad de la República, Uruguay
Rebecca Deneckere — Centre de Recherche en Informatique, France
Chiara Di Francescomarino — University of Trento, Italy
Rodrigo Falcão — Fraunhofer IESE, Germany
Hans-Georg Fill — University of Fribourg, Germany
Andrew Fish — University of Brighton, UK
Agnès Front — Université Grenoble Alpes, France
Ignacio García Rodríguez de Guzmán — University of Castilla-La Mancha, Spain
Sepideh Ghanavati — University of Maine, USA

Mohamad Gharib	University of Tartu, Estonia
Ana-Maria Ghiran	Babeş-Bolyai University of Cluj-Napoca, Romania
Giovanni Giachetti	Universitat Politècnica de València, Spain
Cesar Gonzalez-Perez	Incipit CSIC, Spain
Jaap Gordijn	Vrije Universiteit Amsterdam, The Netherlands
Miguel Goulão	Universidade Nova de Lisboa, Portugal
Martin Henkel	Stockholm University, Sweden
Jennifer Horkoff	University of Gothenburg and Chalmers University of Technology, Sweden
Felix Härer	FHNW University of Applied Sciences and Arts, Switzerland
Mirjana Ivanovic	University of Novi Sad, Serbia
Christos Kalloniatis	University of the Aegean, Greece
Oliver Karras	Leibniz Information Centre for Science and Technology, Germany
Evangelia Kavakli	University of the Aegean, Greece
Manuele Kirsch Pinheiro	Paris 1 Panthéon-Sorbonne University, France
Elena Kornyshova	CNAM, France
Tong Li	Beijing University of Technology, China
Lidia Lopez	Universitat Politècnica de Catalunya, Spain
Beatriz Marin	Universitat Politècnica de València, Spain
Andrea Marrella	Sapienza University of Rome, Italy
Raimundas Matulevicius	University of Tartu, Estonia
Giovanni Meroni	Technical University of Denmark, Denmark
Fredrik Milani	University of Tartu, Estonia
Denisse Muñante	ENSIIE & Télécom SudParis, France
Elena Navarro	University of Castilla-La Mancha, Spain
Kathia Oliveira	Université Polytechnique Hauts-de-France, Spain
Ana Paiva	University of Porto, Portugal
Jose Ignacio Panach Navarrete	Universitat de València, Spain
Oscar Pastor	Universitat Politècnica de València, Spain
Rūta Pirta	Riga Technical University, Latvia
Thomas Polacsek	ONERA, France
Henderik A. Proper	TU Wien, Austria
Ben Roelens	Open Universiteit, The Netherlands
Patricia Rogetzer	University of Twente, The Netherlands
Marcela Ruiz	Zurich University of Applied Sciences, Switzerland
Irina Rychkova	Université Paris 1 Pantheon-Sorbonne, Paris
Rainer Schmidt	Munich University of Applied Sciences, Germany
Florence Sedes	IRIT, Université Toulouse III Paul Sabatier, France

Denis Silveira — Universidade Federal de Pernambunco, Brasil
Anthony Simonofski — Université de Namur, Belgium
Samira Si-Said Cherfi — Conservatoire National des Arts et Métiers, France
Monique Snoeck — KU Leuven, Belgium
Pnina Soffer — University of Haifa, Israel
Eric-Oluf Svee — Stockholm University, Sweden
Daniel Stanley Tan — Open Universiteit, The Netherlands
Ernest Teniente — Universitat Politècnica de Catalunya, Spain
Olivier Teste — IRIT, France
Porfirio Tramontana — University of Naples Federico II, Italy
Nicolas Travers — Léonard de Vinci Pôle Universitaire, France
Gianluigi Viscusi — Linköping University, Sweden
Yves Wautelet — Katholieke Universiteit Leuven, Belgium
Hans Weigand — Tilburg University, The Netherlands

Program Committee, Forum

Judith Barrios Albornoz — University of the Andes, Venezuela
Victoria Döller — University of Vienna, Austria
Riccardo Coppola — Politecnico di Torino, Italy
Georgios Koutsopoulos — Stockholm University, Sweden
Istvan David — McMaster University, Canada
Clara Ayora — Universidad de Castilla-La Mancha, Spain
Michalis Pavlidis — University of Brighton, UK
Abdelaziz Khadraoui — University of Geneva, Switzerland
Christoforos Ntantogian — Ionian University, Greece
Francisca Pérez — Universitat Politècnica de València, Spain
Ben Roelens — Open Universiteit, The Netherlands
Mark Mulder — TEEC2, The Netherlands
Giovanni Meroni — Technical University of Denmark, Denmark
Patricia Martin-Rodilla — Spanish National Research Council, Spain
Dominik Bork — TU Wien, Austria
Elena Kornyshova — CNAM, France
Cinzia Cappiello — Politecnico di Milano, Italy
Sotirios Liaskos — York University, Canada
Giovanni Giachetti — Universitat Politècnica de València, Spain
Yves Wautelet — Katholieke Universiteit Leuven, Belgium
Gianluigi Viscusi — Linköping University, Sweden
Pedro Valderas — Universitat Politècnica de València, Spain
Jose Luis de la Vara — University of Castilla-La Mancha, Spain

Program Committee, Doctoral Consortium

Carina Alves	Universidade Federal de Pernambuco, Brazil
Cristina Cachero	Universidad de Alicante, Spain
Emilio Insfran	Universitat Politècnica de València, Spain
Gustavo Rossi	Universidad Nacional de La Plata, Argentina
Jean-Michel Bruel	University of Toulouse, France
Jolita Ralyté	University of Geneva, Switzerland
Jose Luis de la Vara	University of Castilla-La Mancha, Spain
Maribel Yasmina Santos	University of Minho, Portugal
Miguel Goulão	Universidade Nova de Lisboa, Portugal
Nelly Condori Fernández	Universidad de Santiago de Compostela, Spain
Renata Guizzardi	University of Twente, The Netherlands
Saïd Assar	Institut Mines-Télécom Business School, France
Selmin Nurcan	Université Paris 1 Panthéon - Sorbonne, France
Sergio Firmenich	Universidad of Loyola Andalucía, Spain
Xavier Franch	Universitat Politècnica de Catalunya, Spain

Additional Reviewers

Acitelli, Giacomo
Adiba, Eudes
Baldwin, Wilder
Bono Rossello, Nicolas
Borcard, Daniel
Bromuri, Stefano
Bulmer, Dylan
Casciani, Angelo
Cavalcanti Ribeiro, Ana Beatriz
Chan, Anouck
Curty, Simon
Danthine, Antoine
Gallego, Víctor
Gaspar Vilallba, Alberto
Georgiou, George
Giachetti, Giovanni
Hilvert, Adir
Jokste, Lauma
Jonathan, Gideon Mekonnen
LaChance, Clark
Lektauers, Arnis
Linkevics, Gusts
Macías, Aurora
Matisons, Ralfs
Milani, Fredrik
Mosāns, Guntis
Muff, Fabian
Parshutin, Sergei
Pecerska, Jelena
Pourghasemi Fatideh, Ali
Pretel, Elena
Romanovs, Andrejs
Roponena, Evita
Schinckus, Malik
Zangogianni, Paraskevi-Chrysovalantou

Keynote Talks

The Next 700 Low-Code Platforms

Tijs van der Storm[1,2]

[1] Centrum Wiskunde & Informatica (CWI), Amsterdam, Netherlands
storm@cwi.nl
[2] University of Groningen, Groningen, Netherlands

Abstract. There's too much code in the world and it's growing at staggering exponential rates. No approach in software engineering or programming paradigm has made a dent in this trend. What should we do? Buy a low-code platform? Will AI bring us salvation? The key to successful software engineering is maintaining domain knowledge and architectural knowledge, the "what" and "how". Why not maintain them as code? I argue that the in-house development of domain-specific languages (DSLs), modeling different aspects of a software system, offers a potential way out. State-of-the art language workbenches make this approach feasible, which I demonstrate with a functional low-code platform, developed in under 1000 lines of code. Language engineering for all!

Keywords. Software engineering · domain-specific languages (DSLs) · language engineering · language workbenches · knowledge management

Programming is a form of encrypting requirements in code. Software projects end up as large-scale knitted castles, tangled webs of implementation detail, boilerplate, API appeasement, and through it all, one might find remnants of actual features, scattered and dispersed over thousands if not millions of lines of code. As the code base grows, quality inevitably deteriorates. The ripple effect of changes becomes harder and harder to predict or understand. Too afraid to break stuff in distant parts of the system, developers give up on refactoring. At a certain point in time the only fixes we dare to commit are workarounds...

Complexity is the bane of software development. It's an oft-cited fact that for every 25 percent increase in problem complexity, there is a 100 percent increase in solution complexity. Whether this is literally true or not, I leave to the reader to ponder. What we do know is that software needs to evolve to be alive, and that increase in size hinders maintenance. A recent estimate states that a 2020s hybrid car already contains 150 MLOC, and projects that fully self-driving cars will run on 500 MLOC. There's too much code in the world, and it's increasing at staggering rates.

No programming language or software technology has managed to make a dent in these trends. Yes, we have better, safer, nicer, or faster programming languages, but the economy of code size has largely remained the same. If you squint, all programming languages are the same: different flavors of procedures, expressions, conditionals, and loops. The problem is: general-purpose languages cannot cut corners in terms of expressivity; they have to serve as many clients as possible. The result is that they are mediocre (but functional) at everything, but they excel at nothing.

No software engineering methodology has had any significant effect on programmer productivity or code reduction. Components, frameworks, libraries, patterns, objects, functional programming, or AI—while each of these things has its merits from some point of view, the fact remains that code is all around us, it's not going away, and it's growing. In fact, one could argue that the current hype around AI makes things even worse: AI for programming makes it *much* too easy to produce code *faster*. I will not be surprised if a plot of the global code footprint over time will show a spike when generative AI took the stage...

Another trend is offered by so-called low-code or no-code platforms (e.g., Mendix, Outsystems, Bettyblocks, etc.): often graphical environments which allow citizens to define applications with little or no coding. Indeed, such platforms do increase productivity in certain domains (most prominently business apps), but nevertheless, just like the 4GLs of old, organizations become dependent on such platforms or tools. When a vendor stops support or collapses, what then? The fact that 4GLs were *bought but not owned* caused many companies and organizations to end up with large code bases of unmaintainable legacy code, written in unsupported and obscure languages, with no influx of new generations of programmers. In a sense, one might expect that low-code platforms are paving the way for a 4GL crisis of the future.

What if we could build *our own* low-code platform? So that we own it, so that we understand it fully, so that we can change it to our ever-evolving requirements, wishes, and circumstances? Would it be feasible for ordinary organizations to maintain a set of little languages to define their applications in? In this keynote I argue that maybe this is a way out of the tar pit of complexity, by presenting a radical design experiment: building a low-code platform from scratch, in fewer than 1000 lines of code.

I present the argument that if a single person, using the right tools, can build an operational low-code platform with so little effort (gratuitously taking lines of code as a proxy for effort), wouldn't it be feasible to scale it up, say, a hundred times, to develop a platform suitable for production use? If so, businesses could reap the benefits of increased productivity, without depending on third parties, without paying for features they don't use, without dealing with the constraints of the offered product. In other words, they would own a bespoke low-code platform.

Software engineering where DSLs are the central software development artifact, not designed as a *product*, but as a way to organize the software development process and the knowledge about the problem domain in a traceable, executable manner. Language engineering and design becomes the central activity. Don't buy into a low-code platform—build your own!

Challenges in Forming European Consortia for Ethical and Human-Centered AI: Overcoming Barriers to Collaborative Innovation

Rebeca Gutierrez

Ayesa Advanced Technologies S.A., Seville, Spain
http://rgutierrezs.com/

Abstract. In the context of the increasing complexity of the information systems that surround us and the intensified competition in the generation of innovative ideas, collaboration around European values and distinctive approaches emerges as a key catalyst to achieve significant advances in artificial intelligence, overcoming obstacles and cultivating a high-impact innovation ecosystem for the transformation of companies, entities and where high-impact actions are obtained in the citizenship. This talk will analyse strategies that strengthen cooperation between academia, industry and the public sector, with a particular focus on flagship initiatives such as Horizon Europe, CINEA, InnoFund or the recently launched European Network of AI Centres of Excellence and AI Factories. The discussion will focus on how these collaborations facilitate access to critical resources, foster efficient knowledge transfer and promote agile adoption and development of AI technologies. In parallel, the conference will address the ethical and regulatory challenges inherent in collaborative AI innovation by exploring regulatory frameworks that encourage the development of responsible and trustworthy AI, without compromising creative and innovative potential. It will also examine open innovation models designed to boost European competitiveness in the global AI arena, balancing the need for technological advancement with ethical and societal imperatives.

Keywords. Artificial Intelligence · Innovation · European Collaboration · AI Ethics · Human-Centered AI

References

1. European Commission Homepage. https://commission.europa.eu/
2. AI Factories. https://digital-strategy.ec.europa.eu/en/policies/ai-factories
3. AI Europe Collaborations. https://digital-strategy.ec.europa.eu/en/news/second-wave-ai-factories-set-drive-eu-wide-innovation
4. AI Literacy. https://digital-strategy.ec.europa.eu/en/events/third-ai-pact-webinar-ai-literacy
5. AI Act. https://digital-strategy.ec.europa.eu/en/policies/regulatory-framework-ai

Explanation, Semantics, Ontology and Trustworthiness

Giancarlo Guizzardi◉

Semantics, Cybersecurity & Services (SCS), University of Twente, The Netherlands
g.guizzardi@utwente.nl

Abstract. Cyber-human systems are formed by the coordinated interaction of human and computational components. The latter are justified to the extent that they are meaningful to humans, in both senses of 'meaning', that is, in the sense of *semantics* and in the sense of *purpose* (or *significance*). On the one hand, the data these components manipulate only acquire meaning when mapped to shared human conceptualizations of the world; on the other hand, they can only be justified if ethically designed. Cyber-human systems are trustworthy if the interoperation of their components is meaning preserving, i.e., if we can semantically interoperate these components; and transparently demonstrate (i.e., explain) how their interoperation positively contributes to human values and goals. In this talk, I will present a notion of explanation termed *Ontological Unpacking*, which aims at explaining symbolic domain descriptions (e.g., conceptual models, knowledge graphs, logical specifications). I show that it is this explanatory nature that is required for semantic interoperability, and hence trustworthiness. Finally, I will argue that the current trend in XAI (Explainable AI) in which 'to explain is to produce a symbolic artifact' is an incomplete project, as these artifacts are not 'inherently interpretable', and that they should be taken as the beginning of the road to explanation, not the end. The talk is strongly based on [1, 2, 3, 4].

Keywords. Real-world semantics · Ontology explanation · Ontological unpacking · Semantic interoperability · Trustworthiness

References

1. Confalonieri, R., Guizzardi, G.: Explanation, semantics, and ontology. Data Knowl. Eng. **153** (2024)
2. Guizzardi, G., Guarino, N.: On the multiple roles of ontologies in explanations for neuro-symbolic AI. Neurosymbolic Artif. Intell. **1** (2025)
3. Bernasconi, A., Guizzardi, G., Pastor, O., Storey, V.C.: Semantic interoperability: ontological unpacking of a viral conceptual model. BMC Bioinform. **23**(1) (2022)
4. García, A., Bernasconi, A., Guizzardi, G., Pastor, O., Storey, V. C., Panach, I.: Assessing the value of ontologically unpacking a conceptual model for human genomics. Inf. Syst. **118** (2023)

Contents – Part II

Machine-Learning and Generative AI Applications

Can Generative AI Mitigate Strategic Decision-Making Complexity?
An Empirical Exploration of RAG-Augmented Decision-Support Systems 3
   Jordan Abras, Corentin Burnay, and Stéphane Faulkner

Cross-Lingual Entity Linking Using GPT Models in Radiology Abstracts 20
   Mariana Dias and Carla Teixeira Lopes

Modeling Out-of-Vocabulary Words via Grammatical Fusion 38
   Dror Mughaz

Early Length of Stay Prediction at Admission in Short-Stay Hospitals 52
   Mohamed Gharbi, Christine Verdier, Maria Di Mascolo,
   and Jean-Marc Babouchkine

Alignment of Schema-Only and Instance-Only Data Sources Using Large
Language Models 67
   Nour Elhouda Kired, Franck Ravat, Jiefu Song, and Olivier Teste

RCIS Forum

Can Llama 3 Accurately Assess Readability? A Comparative Study Using
Lead Sections from Wikipedia 89
   José Frederico Rodrigues, Henrique Lopes Cardoso,
   and Carla Teixeira Lopes

Research Challenges in Routine Optimization for Synthesizing Software
Robots 98
   J. L. Alonso-Rocha, A. Martínez-Rojas, A. Jiménez-Ramírez,
   and J. G. Enríquez

Mining for Meaning: Ontology-Aware Process Mining Methods Through
Knowledge Patterns 109
   Riley Moher and Michael Gruninger

How to Use a FEM Model as a Basis for Strategic-Level Risk Analysis 120
  Toomas Saarsen and Ilia Bider

Modelling Neural Network Models 130
  Nadia Daoudi, Ivan Alfonso, and Jordi Cabot

Quantifying the Magnitude of Violation: Predictive Compliance
Monitoring Approaches 140
  Qian Chen, Stefanie Rinderle-Ma, and Lijie Wen

Evaluating Programming Optimization Techniques in C and Python:
Impact on Energy Consumption 151
  *Carlos Pulido, Félix O. García, M¹ Ángeles Moraga,
  Coral Calero, Miguel Baños-González, Jorge Cancho-Casado,
  and Javier Corral-García*

An Analysis of Resilience in Digital Business Ecosystems 162
  Beāte Krauze

AI Auditing: Towards a Practicable Model 172
  A. B. van Wingerden and H. Weigand

From Acquiring to Suggesting DL Design Choices with Agility: A System
Design 183
  *Gustavo Rodrigues dos Reis, Mario Cortes Cornax, Adrian Mos,
  and Cyril Labbé*

RCIS Doctoral Consortium

Digital Twins for Incident Detection and Response 197
  Konstantinos E. Kampourakis

Towards an Enterprise Architecture Based Approach for the Development
of Digital Twins for Sustainable Real Estate Management 207
  Marianne Schnellmann

A Method for Domain Reference Model Inference Through Knowledge
and Data Intelligent Unifiers 218
  Pedro Guimarães

Tutorials

Problematising and Ideating for Design Science Research 231
   John R. Venable

Taking Control of Our Privacy When Using Mobile Apps: Checks and Tools ... 232
   M. Mercedes Martínez-González, Alejandro Pérez-Fuente, Amador Aparicio, and Pablo-Abel Criado-Lozano

Introduction to Fractal Enterprise Model (FEM) and FEM Toolkit 234
   Ilia Bider

Author Index 237

Contents – Part I

Information Systems Quality

TRACE4PM: Trace Related Analysis and ClustEring for Process
Modeling of Users' Interactions in Information Systems 3
    *Marwa Trabelsi, Noura Joudieh, Amira Ania Dahache, Cyrille Suire,
and Ronan Champagnat*

Challenges in Data Quality Management for IoT-Enhanced Event Logs 20
    *Yannis Bertrand, Alexander Schultheis, Lukas Malburg, Joscha Grüger,
Estefanía Serral Asensio, and Ralph Bergmann*

Blockchain-Based Trust Management System for Enhancing Security
in SIoT 37
    Raouf Jmal, Mariam Masmoudi, Ikram Amous, and Florence Sèdes

A Comprehensive Review on Equivalent Mutant Detection Using Machine
Learning 52
    *Gabriel Guerrero-Contreras, Sara Balderas-Díaz,
Pedro Delgado-Pérez, and Inmaculada Medina-Bulo*

Security, Risk and Strategy

Identifying and Detecting Patterns in Work Organization with Active
Window Tracking 71
    *Mari A. J. Braakman, Iris Beerepoot, Maria Peeters, Eva Knies,
and Hajo A. Reijers*

Black Swan Theory for Navigating Trust in Mixed-Traffic Environments 87
    Hind Bangui, Barbora Buhnova, and Mouzhi Ge

Friend, Foe, or Target? Domain Models as Risk Deterrents, Risk Sources,
and Assets at Risk 103
    *Isadora Valle, Tiago Prince Sales, Eduardo Guerra, Ítalo Oliveira,
Renata Guizzardi, Luiz Olavo Bonino da Silva Santos,
Henderik Proper, and Giancarlo Guizzardi*

Conceptual Modelling and Ontologies

Towards an Ontology of Type-Level Phenomena for System Modeling 121
  *Rodrigo Fernandes Calhau, João Paulo A. Almeida,
  Giancarlo Guizzardi, Raquel Hoffmann, Luís Ferreira Pires,
  James Logan, and João Rebelo Moreira*

Low-Code Browser Front-End Automation Using RDF Graphs
and a Domain-Specific Language for UX Representation 140
  Ştefan Uifălean and Robert Andrei Buchmann

Risk Response: An Adversarial Discourse Game and Group Modelling Tool ... 156
  Max Willis, Greta Adamo, and Anna Sperotto

Modeling Methods and Requirements Engineering

Modelling Hierarchies of Organisational Rules 175
  Jöran Lindeberg, Martin Henkel, and Katarina Fast Lappalainen

If Complexity Is the Problem, Collaboration Is the Solution: Drivers
and Trust-Building Interventions in DSR 192
  Xabier Garmendia and Oscar Díaz

Synthline: A Product Line Approach for Synthetic Requirements
Engineering Data Generation Using Large Language Models 208
  Abdelkarim El-Hajjami and Camille Salinesi

The Role of Ethics in Requirements Engineering for Developing
Information Systems 226
  *Christopher Julian Kern, Karin Hübner, Leo Poss, Stefan Schönig,
  and Julia Kroenung*

Databases and Information Management

Model-Agnostic Evolution Management 245
  Pavel Koupil, Jáchym Bártík, and Irena Holubová

A Flexible Framework for Transposition-Aware Querying of a Musical
Score Database 262
  Adel Aly, Olivier Pivert, and Virginie Thion

Implementing Digital Twin Query Views 279
  Emilio Carrión and Pedro Valderas

Framework for the Design and Development of an Automated Virtual
Data Integration System for Genomic Data 295
   Ana León Palacio

Human Factors in Information Systems

Developing a Design Features Taxonomy of Human-Computer Interaction
in Social Media that Affect User Engagement and Addictive Behaviors 313
   *Maria Fernanda Granda, María-Belén Sarmiento,
   Ana-Gabriela Nuñez, Ricardo Maldonado, and Otto Parra*

User Correction of Misinformation on Social Media: Perceived and Actual
Social Norms 331
   Selin Gurgun, Emily Arden-Close, Keith Phalp, and Raian Ali

Do Social Media Simultaneously Contribute to Well-Being and Use
Disorder? Empirical Evidence and Design Challenges 347
   *Tourjana Islam Supti, Ala Yankouskaya, Sameha Alshakhsi,
   Areej Babiker, Dena Al-Thani, and Raian Ali*

Business Process Engineering and Management

Multi-perspective Next Event Prediction in PPM via Heterogeneous Graph
Neural Networks 365
   *Sebastiano Dissegna, Chiara Di Francescomarino,
   and Massimiliano Ronzani*

OCPQ: Object-Centric Process Querying and Constraints 383
   Aaron Küsters and Wil M. P. van der Aalst

Investigating Tailored Retraining for Online Process Predictions Using
Log Features 401
   Suhwan Lee, Xixi Lu, and Hajo A. Reijers

Navigating the Challenges of Process Mining Use Case Definition:
A Qualitative Study 418
   Kerstin Haug, Pol Schumacher, Holger Wittges, and Stefanie Rinderle-Ma

Zero-Shot Approaches for the Extraction of Event Logs from Medical Notes ... 435
   Allmin Susaiyah and Natalia Sidorova

Review of Design of Business Process Simulation Models 452
   Samira Khraiwesh and Luise Pufahl

Exploring Webcam Eye Tracking Software for Robotic Process
Automation: A Pilot Benchmarking Study 470
  Manuel García-Romero, Antonio Martínez-Rojas,
  José González Enríquez, and Andrés Jiménez-Ramírez

Author Index 487

Machine-Learning and Generative AI Applications

Can Generative AI Mitigate Strategic Decision-Making Complexity? An Empirical Exploration of RAG-Augmented Decision-Support Systems

Jordan Abras[✉], Corentin Burnay, and Stéphane Faulkner

Namur Digital Institute, University of Namur, Namur, Belgium
{jordan.abras,corentin.burnay,stephane.faulkner}@unamur.be

Abstract. Strategic decision-making (SDM) is inherently complex, requiring decision-makers to balance short-term constraints with long-term objectives while navigating vast, often overwhelming, volumes of data. These challenges are further exacerbated by the cognitive limitations of human decision-makers. In this study, we investigate the integration of Retrieval-Augmented Generation-enhanced Decision-Support Systems (RAG-DSS) – a technology combining Generative AI capabilities with external knowledge retrieval – into SDM processes to mitigate perceived complexity and support strategic decision tasks through natural language interaction. Employing a case study methodology, we investigate a business management simulation seminar, where 12 participants engaged with a context-specific RAG-DSS. Data collection included analysis of participants' interactions with the system (prompts) and semi-structured interviews conducted before and after its introduction. The findings demonstrate that the RAG-DSS reduces perceived SDM complexity by facilitating faster and more intuitive access to relevant, structured information, generating context-specific strategic suggestions, and enabling predictive analyses and simulations using natural language queries. Participants reported increased confidence and a smoother decision-making process, with the RAG-DSS serving as a tool for reassurance and validation. However, challenges emerged, including risks of iterative user-chatbot loops and concerns about over-reliance on the system's outputs. This research highlights the potential of RAG-enhanced DSSs to address SDM complexity, offering actionable insights for their integration into organizational decision processes while identifying areas for further research to optimize their implementation and mitigate unintended consequences.

Keywords: Generative AI · Decisions-Support Systems · Strategic Decision-Making · Complex Decision Process · Large Language Model · Retrieval Augmented Generation

1 Introduction

Decision-making lies at the heart of any organizational activity, ranging from routine operational decisions to critical, long-term strategic decisions [12]. Strategic decisions, in particular, are critical as they significantly influence the long-term success, sustainability, and overall course of organizations [5]. Despite their critical importance, strategic decision-making (SDM) remains a complex process.

SDM complexity stems from the necessity to align long-term goals with short-term constraints while navigating high levels of uncertainty and ambiguity in the decision environment. This complexity demands that decision-makers base their analyses and decisions on all available information. However, in today's data-rich contexts, they face the additional challenge of processing vast volumes of information that must be filtered, interpreted, and synthesized to reach well-founded conclusions. While some of this information is structured and easily accessible, much remains unstructured and difficult to capture, adding yet another layer of challenge to the process [5]. Furthermore, the cognitive capacities of human decision-makers are inherently limited, which in turn hinders their ability to fully process and utilize this wealth of information effectively [13,18].

To overcome this challenge of bounded capacities in an increasingly data-intensive and complex environment, data-driven decision-support systems (referred to as "DSSs" from here on for simplicity and clarity) were introduced in the 1990s as a means of providing an easier access to centralized, structured sources of information to support decision-makers [2,15]. Advances in computing power and artificial intelligence (AI) have since transformed DSSs into sophisticated tools capable of executing complex data analysis, simulations, and predictive modeling [1]. However, these enhancements have led to a paradox: The ever-increasing volume of data to be analyzed, combined with ever-improving computing capabilities, has led to systems so rich in information that decision-makers are often overwhelmed, facing what is commonly referred to as "information overload" [4]. This paradox raises a critical question: How can we continue to enhance the computational power and information richness of DSSs to support SDM, while mitigating the adverse effects of information overload to promote more efficient yet simpler decision processes?

One emerging technology that holds significant promise in this regard is Generative AI (GenAI). GenAI technologies extend beyond the capabilities of traditional, deductive, AI systems by demonstrating the ability to generate original content – such as text, images, videos, or code – based on large training datasets and user interactions [6]. Among the various categories of GenAI, Large Language Models (LLMs) may offer particularly relevant features for supporting strategic decisions. LLMs are advanced AI systems trained on vast amounts of textual data, enabling them to understand and respond to natural language queries with coherent and contextually relevant outputs. By leveraging Retrieval-Augmented Generation (RAG), LLMs can be further enhanced with external datasets in various formats, significantly improving their capacity to deliver more precise, context-specific responses. This augmentation allows LLMs to transcend their original training data, offering more accurate and tailored insights based on real-time or domain-specific information [17]. In this study, we focus on the integration of RAG-enhanced LLMs within DSSs to create a new generation of DSSs (we call RAG-DSS) that enable decision-makers to interact with the system using natural language. By facilitating intuitive access to tailored insights, interpreting complex data, and generating actionable recommendations, RAG-DSSs have the potential to address the cognitive and informational challenges associated with SDM. As a result, this study explores the following research question: "Does the introduction of a RAG-enhanced DSS help mitigate the complexity of strategic decision-making processes, and if so, how is this achieved?".

This preliminary study is part of a broader research project examining the role of GenAI-enhanced DSSs in strategic management and decision science. Employing a case-study approach [20], we conducted research within a business management simulation seminar where 12 participants were provided with a RAG-DSS to support their SDM tasks. Through the collection of 344 prompts – representing user interactions with the system – and three series of semi-structured interviews, we analyzed how participants engaged with the tool, integrated its outputs into their decision-making processes, and perceived its impact on SDM complexity. This study identifies usage patterns and explores how RAG-DSSs influence decision-making, providing insights into their potential to reduce complexity in data-intensive environments.

2 Background

Strategic decisions are characterized by their inherent complexity, as they are made in environments fraught with uncertainty, ambiguity, and unpredictability. Such decisions are often made in contexts where perfect information is unattainable and the analytical capabilities of decision-makers are constrained by cognitive limitations. Even when supported by AI technologies like predictive Bayesian networks or machine learning models, human decision-makers operate within the limits of their cognitive capacities, which hampers their ability to make fully rational decisions [13]. Consequently, decision-makers often operate under Simon's "bounded rationality" [16], making decisions based on satisfactory rather than optimal solutions that meet minimum acceptability thresholds.

The process of SDM in environments characterized by complexity and uncertainty involves multiple interconnected activities, that researchers have long sought to model and optimize. The foundational model for SDM guiding this study, as shown in Fig. 1, synthesizes influential frameworks, including Trunk et al. [18], who present a comprehensive model for organizational decision-making under uncertainty, alongside the seminal work of pioneers in decision theory under uncertainty, such as Simon [16] and Mintzberg [13]. This model has been adapted and streamlined to fit our study's focus on a data-intensive environment.

The process initiates with problem recognition and goal definition, where decision-makers identify the need for a decision and clarify the objectives that the decision should achieve. Following this, decision-makers engage in information gathering, identifying relevant data that aligns with their specific decision context. The gathered information then undergoes analysis to generate actionable insights, which support the development of alternative solutions. Alternatives are then evaluated against predefined goals, with the final decision emerging from the most satisfactory option.

Although this model appears sequential, SDM is rarely a linear process. The iterative nature of SDM, as observed by Mintzberg [13] and Trunk [18], incorporates numerous feedback loops. This iterative cycle emerges from the continuous influx of new information and evolving analyses, which can prompt adjustments in the decision goal or the development of new alternatives even at advanced stages, such as during the evaluation phase.

As exposed in the introduction, today's DSSs can be so computationally powerful, offering access to vast amounts of information, that it can easily overwhelm decision-makers who struggle to filter and prioritize the influx of data. This challenge is particularly evident in the information identification and analysis phases, where the limits of human cognitive and analytical capacity can hinder comprehensive insight generation. As a result, high cognitive demands are placed on decision-makers, complicating the process of alternative development and, ultimately, the evaluation and selection of feasible options.

This study aims to examine the impact of integrating a RAG-enhanced LLM component within a DSS to ease the SDM process. Specifically, we explore whether RAG-DSSs can alleviate decision complexity by effectively supporting the decision phases depicted in Fig. 1, and if so, how.

3 Methodology

Given the exploratory nature of this research, we adopted a qualitative approach to gain insights into the phenomenon under study within its natural context, as is recommended for novel and under-researched areas. To this end, we employed semi-structured interviews as qualitative methods to investigate the impact of introducing a RAG-DSS in the SDM process. Through this qualitative lens, we aimed to uncover patterns regarding how decision-makers interact with the conversational component and whether the inclusion of this component contributes to a perceived reduction in decision complexity. Furthermore, we gathered and analyzed all prompts generated by the participants to examine and uncover usage patterns in their interactions and queries.

Methodologically, we employed a case study approach, as articulated by Yin [20]. The rationale for selecting the case study methodology lies in its appropri-

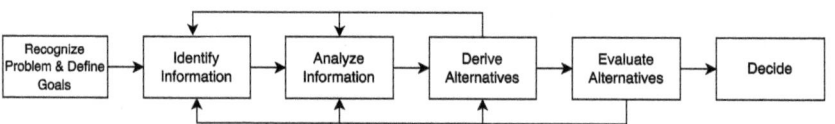

Fig. 1. Simplified SDM process in a data-intensive environment. This process is derived from research from [13, 14, 16, 18].

ateness for studying novel, contemporary phenomena within real-life contexts. Given that GenAI technologies are not yet widely integrated into DSSs, a case study design offers the most effective framework for exploring their potential impact. This approach allows us to examine the dynamics of RAG-enhanced LLM integration into SDM processes and observe how decision-makers interact with this new technology in a practical, data-driven scenarios.

This case study is part of an ongoing broader multi-site research design exploring GenAI's role in SDM. Future cases will be used to compare findings and assess their relevance, generalizability, and theoretical contributions. This iterative multi-site approach aims to identify broader patterns and develop a cohesive framework capturing GenAI's impact on SDM processes.

3.1 Case Description

This study was conducted during a business management simulation seminar designed to immerse participants in executive decision-making. Participants assumed C-suite roles (e.g., Chief Operating Officer, Chief Financial Officer) and managed a fictional company over the course of a semester. Each week, they made strategic decisions on key organizational variables, including product pricing, workforce adjustments, capital investments, and production levels. These decisions were processed by a simulator replicating real-world market dynamics, competitive pressures, and organizational behavior, generating detailed reports on economic, financial, operational, and social outcomes. Developed through an international collaboration among researchers from universities in Belgium, Canada, and Japan, the simulator also enabled interactions with key managerial stakeholders through a dedicated platform. At the end of the seminar, domain experts in relevant management fields systematically evaluated the final state of the company and the set of decisions of the participants.

The sample consisted of 12 participants with diverse expertise and experience levels across various industries, all of whom voluntarily agreed to take part in the research. Their backgrounds offered a wide array of perspectives on organizational decision-making and technological experiences, thereby enriching the study. Some demographic information is provided in Table 1.

3.2 Data Collection

Data collection was carried out through a multi-phase approach to ensure a comprehensive understanding of participants' decision-making processes before and after the introduction of the RAG-DSS. The data collection protocol unfolded in three distinct stages:

Phase 1: Baseline Decision-Making (Without LLM Support). In the first stage, participants completed four rounds of strategic decisions using only the simulator's generated reports as information source. These reports provided traditional decision-making indicators, reflecting various economic, financial, operational, and social metrics generated by the simulator. No additional decision-support tools were made available during this phase.

Following this baseline phase, we conducted the first round of semi-structured interviews. These interviews focused on capturing participants' perceptions of the complexity inherent in the decision-making process and their reliance on other decision-support tools, if any, during this period. The aim of this stage was to establish a baseline understanding of the participants' SDM process and behavior and to identify the challenges they faced before the introduction of the LLM component.

Phase 2: RAG-DSS Augmented SDM. In the second phase, participants were introduced to a RAG-DSS, designed to complement the traditional reports they had been using. This system took the form of a chatbot, utilizing OpenAI's GPT-4 model, augmented with historical data from the seminar through the technique of RAG. RAG allows generalist models, such as GPT-4, to enhance their knowledge by accessing specific external datasets [11]. In this study, the chatbot was augmented with the seminar's rules, historical data on decisions and their results from past rounds of decision-making, and typical performance indicators (financial, operational, and social). Additionally, the chatbot had access to context-specific information regarding the fictional actors and organizations involved in the seminar, offering participants a richer and more contextualized source of decision support.

Table 1. Demographics of the study participants (in frequency).

Category	Value	Count
Age	18–29 (Young Adults)	5
	30–44 (Adults)	5
	45–59 (Mature Adults)	2
Gender	Male	7
	Female	5
Education Level	Bachelor's Degree/Undergraduate	9
	Master's Degree	2
	PhD Degree	1
Industry	Insurance & Financial Sector	3
	Healthcare	1
	Public Services	3
	Technology & Engineering	2
	Human Resources Management	1
	No Disclosure	2

Participants then completed four additional rounds of decision-making, this time with the option to interact with the chatbot for assistance, advice, and answers to queries related to their strategic decisions. After this phase, we conducted a second set of interviews, focusing on participants' experiences with the RAG-DSS. Specifically, the interviews explored how the chatbot influenced their decision-making processes and their overall perceptions of RAG-DSS-supported decision-making complexity.

Phase 3: Comparative Reflection and Final Interviews. At the conclusion of the seminar, we conducted a final round of interviews to allow participants to reflect on the two experienced phases of the study. The interviews aimed to gather insights into participants' perceptions of the relative complexity of the decision-making processes across both phases, their evaluation of the RAG-DSS's utility, and their final reflections on the potential of RAG-enhanced LLM technology for strategic decision-support.

Throughout the study, we also collected participants' prompts, with their consent, to gain additional insights and further enrich our analyses on their usage patterns of the RAG-DSS. In the end, we collected 12 h of interviews and 344 queries submitted to the RAG-DSS.

3.3 Data Analysis

All interviews were progressively transcribed and anonymized, and the prompt logs from the chatbot interactions were also extracted and anonymized for analysis. The interviews were inductively analyzed using NVivo, following a systematic thematic analysis approach that adhered to the guidelines of Gioia et al. [8]. We initially employed open coding to extract first-order concepts directly from the participants' narratives. Subsequently, we conducted axial coding to synthesize these concepts into second-order themes, exploring the relationships and patterns among them. Finally, we utilized selective coding to distill the second-order themes into aggregate dimensions, resulting in a cohesive narrative that encapsulates the integration of the RAG-DSS into the participants' SDM process.

The prompts were analyzed in terms of both content and frequency, with a focus on the types of requests made to the chatbot and their correspondence to various phases of the decision-making process. The coding and prompt analysis processes were collaboratively reviewed by all co-authors at each step, ensuring consistency, rigor, and high-quality outcomes through iterative discussions and collective refinement.

4 Empirical Results

4.1 The SDM Process Experienced by Participants

The initial phase of this research aimed to clarify the SDM process as experienced by participants in the business management seminar and to check its compliance with how the literature defines a SDM process (Fig. 1). To do so, we integrated narrative analysis [10] – drawing from participants' descriptions of their decision-making process – with process mapping to identify and outline the specific steps they followed from one round of SDM to the next. This approach provided a detailed breakdown of the sequential stages in their SDM process, revealing the structured sequence presented in Fig. 2, which has been mapped onto the high-level phases derived from the literature and detailed in Fig. 1.

The SDM process experienced by participants aligns closely with models established in the literature [13,14,16,18]. Participants described beginning with a diagnosis of their company's current state, followed by the definition of goals for the next decision period. They then isolated relevant information from the extensive data available, focusing on key areas such as financial performance, market analysis, and competitor activities.

Subsequently, participants moved to the analysis phase, deriving insights from reports generated by the simulator and feedbacks from the different actors, calculating new indicators (e.g., financial ratios, return on investment) and conducting simulations to anticipate market conditions, including demand and productivity. Based on these analyses, participants formulated pre-decisions, which were collaboratively evaluated through further simulations or computations. The process concluded with participants either reaching a consensus on the most viable decision and finalizing their decisions or agreeing that no solution was sufficiently good and going back to previous steps in the decision process. This structured approach aligns with the iterative nature of SDM described by early researchers [13] and reaffirmed by more recent studies [18].

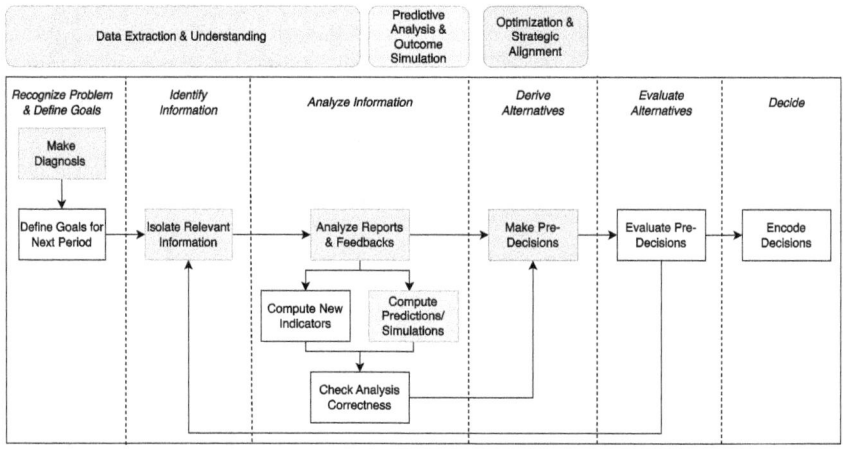

Fig. 2. SDM process mapped from participants' narratives, with the three identified dimensions of SDM complexity highlighted. Each dimension is color-coded and mapped to the affected tasks within the SDM process to illustrate their impact.

Although no LLM or AI component was formally introduced in the study, some participants mentioned using basic AI models to assist with simulations or predictions. Additionally, two participants specifically referred to leveraging GenAI tools such as ChatGPT, Copilot, and Perplexity for strategic advice and evaluating the feasibility and viability of their decisions, while others relied solely on the reports provided by the simulator as their primary DSS. This demonstrates that the technological background of our sample is quite diverse.

4.2 Complexities Experienced in SDM

When describing their decision process, participants were asked to identify tasks they found complex or factors contributing to the complexity of reaching a decision. We analyzed their responses, which led to the identification of three dimensions of complexity within the SDM process, as illustrated in Fig. 2. These dimensions were mapped onto the participants' SDM process to highlight their impact on specific decision-making tasks.

The first dimension, **Data Extraction & Understanding**, encompasses identifying relevant information within extensive data sources, interpreting its significance, and deriving actionable insights from its analysis. Participants often expressed difficulty in isolating meaningful data from overwhelming volumes, particularly when dealing with unfamiliar information. They also noted challenges in understanding the interdependencies among decision parameters.

The second dimension, **Optimization & Strategic Alignment**, highlights the challenges of optimizing decision components while aligning them with available resources and long-term strategic goals. Participants noted difficulties in building optimization models due to the complex interplay of variables and their impacts, as well as in ensuring that detailed management decisions align with overarching strategic objectives.

The third dimension, **Predictive Analysis & Outcome Simulation**, involves the challenges of anticipating decision impacts and simulating potential outcomes in contexts of imperfect information. Participants reported difficulties in predicting the full scope of their decisions' effects, including unintended consequences on other organizational units. They also struggled to account for external variables, such as demand fluctuations and market share volatility, which hindered their ability to make fully informed decisions.

These three dimensions of perceived complexity align closely with challenges highlighted in Sect. 2 concerning decision-making under uncertainty in complex, data-rich environments. The first dimension corresponds to the issue of information overload, emphasizing the significant difficulty human decision-makers face in filtering, isolating, and analyzing relevant information from large datasets to derive actionable insights.

The second and third dimensions are rooted in the concept of bounded rationality, which reflects the cognitive and analytical constraints that limit human decision-makers' ability to achieve fully optimized strategic outcomes. Instead, they often adopt a satisficing approach, selecting alternatives that are acceptable and offer favorable outcomes within the constraints they face. These dimensions also highlight the difficulty in conducting accurate predictions and simulations, a critical task in strategic contexts. Together, they underscore the intricate interplay between information complexity, human cognitive limitations, and the decision-making environment, presenting significant barriers to achieving truly rational and optimized decisions.

These three dimensions of complexity will serve as a structured framework for examining the extent to which the RAG-DSS mitigate the inherent challenges of complexity in the SDM process within our case study.

4.3 Utilization Patterns of the RAG-DSS

In the second phase of this study, participants were granted access to a custom-built, context-specific chatbot integrated as an LLM component augmenting the existing DSS provided by the simulator. At the conclusion of the simulation, all user interactions with the chatbot were collected and analyzed to uncover patterns in usage, including how participants engaged with the chatbot and the types of requests they made.

A total of 344 user-initiated prompts were gathered, of which 188 were directly related to SDM tasks. The remaining prompts encompassed a variety of uses, such as requests for rewriting content, advice on writing styles, and other general queries unrelated to decision-making. The SDM-related prompts were systematically classified based on the decision tasks they supported, resulting in the visualization presented in Fig. 3.

As illustrated in Fig. 3, the majority of the prompts issued by participants were concentrated in the "Identify" and "Analyze" phases of the SDM process (second and third quadrants of Fig. 3), collectively accounting for 72% of all prompts. This observation highlights that the primary use of the LLM component within the DSS seems to be to identify, isolate, retrieve, and analyze data – activities that serve as the foundation for subsequent decision-making phases, which were comparatively less supported by the chatbot. Notably, tasks associated with the complexity dimension "Data Extraction & Understanding" (see Fig. 2) encompassed 113 out of the 188 relevant prompts, representing approximately 60% of the total. This finding underscores the chatbot's predominant role as a data retrieval and analysis tool, addressing challenges related to this dimension of complexity.

Within the "Analyze" phase, another significant subset of prompts focused on tasks related to "Predictive Analysis & Outcome Simulation", a second dimensions of complexity (Fig. 2) totaling 21 prompts (approximately 11%). These interactions demonstrated that participants also engaged with the chatbot to perform simulations and generate predictions, demonstrating a second, relatively lowly exploited, utility as a "simulator" itself.

A third prominent trend in prompt usage emerged during the "Pre-Decision Making" task, which pertains to the "Derive Alternatives" quadrant of the SDM process. This phase accounted for 32 prompts, or 17% of the total relevant prompts. Participants frequently sought suggestions from the chatbot on potential decisions to make, based on strategies, constraints, and objectives, reflecting its relevance to tasks linked to the third complexity dimension "Optimization & Strategic Alignment".

This exploratory analysis of participant prompts reveals that the complexity dimension most addressed by the chatbot in terms of prompt density is "Data Extraction & Understanding", accounting for 60% of the relevant prompts. Within this dimension, 92% of prompts focused on isolating and analyzing relevant data and information. The second most addressed dimension, "Optimization & Strategic Alignment", constituted 17% of the prompts, where participants leveraged the chatbot to obtain strategic suggestions and optimize decisions, particularly in the face of constraints or overlooked considerations. Finally, the "Predictive Analysis & Outcome Simulation" dimension represented 11% of the prompts, demonstrating a relatively low participants' reliance on the chatbot to perform simulations and predictions.

Overall, while "Data Extraction & Understanding" emerged as the most impacted dimension, all three dimensions of complexity were addressed to varying extents through the chatbot's usage patterns. These findings provide a foundational understanding of how a RAG-DSS can support different aspects of SDM complexity. The subsequent phase of analysis, based on participant interviews, will further enrich these observations by exploring perceptions of the chatbot's impact on SDM complexity and its role in the SDM process.

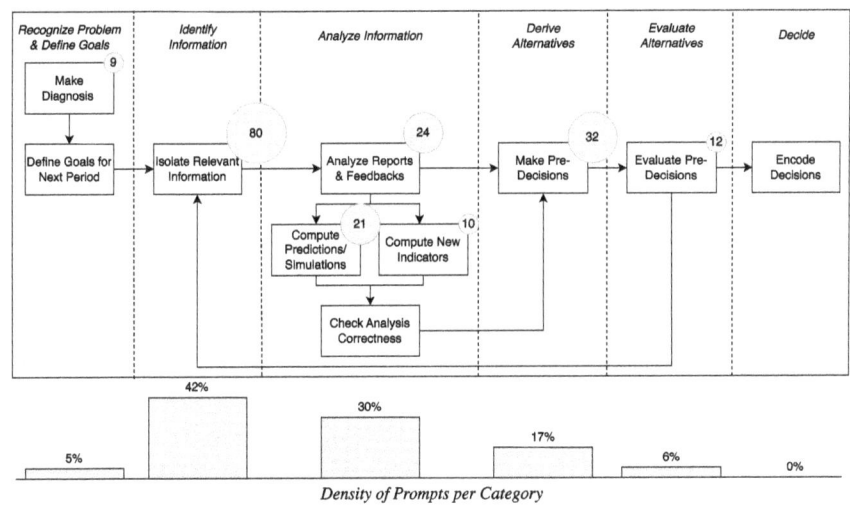

Fig. 3. SDM process annotated with the total number of participant requests corresponding to each decision task. Below the process, the percentage distribution of requests is depicted for each theoretical category of SDM activity, as derived from the process in Fig. 1.

4.4 Perceived Influence of the Chatbot on Strategic Decision-Making Complexity

During the second phase of this research, we conducted interviews with participants to delve deeper into their use of the chatbot and its perceived impact on the complexity of their SDM process. The findings reinforced the insights gained from the prompt analysis and highlighted the nuanced ways the chatbot influenced decision-making complexity.

A primary conclusion is that participants universally experienced a reduction in both the complexity and time required for their SDM process. This reduction was most pronounced in the dimension of "Data Extraction & Understanding". Participants consistently praised the chatbot's ability to streamline access to relevant information, often anticipating their needs and delivering tailored responses in a structured and concise format. They emphasized how this feature reduced the need to navigate through extensive dashboards or dense spreadsheets. Moreover, the chatbot's ability to dynamically adjust the granularity of information based on natural language queries was frequently highlighted as a game-changer, allowing decision-makers to focus on actionable insights rather than data wrangling. The capability of the chatbot to interpret complex data and provide diagnostic summaries of organizational states further accelerated the diagnosis phase, enabling participants to swiftly identify problems or capitalize on opportunities. This not only simplified the process but also enhanced participants' confidence in the accuracy and comprehensiveness of their analyses.

The other two dimensions of complexity also benefited from the chatbot, albeit to a lesser extent. For the dimension of "Optimization & Strategic Alignment", participants valued the chatbot's ability to provide strategic advice and refine their decision alternatives. They often used it to validate their ideas and ensure decisions were systematically aligned with the organization's overarching goals. By acting as both a sounding board and an optimization tool, the chatbot facilitated a more systematic and confident decision-making process. This alignment between short-term actions and long-term strategies was noted as a significant contributor to reducing perceived complexity in this phase of SDM.

The dimension of "Predictive Analysis & Outcome Simulation", while positively impacted, showed a more limited influence. Participants appreciated the chatbot's ability to generate simulations and provide feedback on potential strategies. However, they noted a preference for specialized data analytics tools for performing complex numerical modeling and simulations, which they perceived as more reliable and precise for these tasks. This suggests that while the chatbot offered complementary support, it did not replace established tools for this dimension of complexity.

Beyond reducing complexity, the chatbot was also described as a **reassuring and confidence-enhancing tool**, particularly for participants facing unfamiliar or high-stakes decisions. Its ability to act as an accessible, real-time assistant fostered a sense of support, encouraging decision-makers to explore alternative strategies and challenge their own assumptions. This emotional and cognitive reassurance emerged as an unexpected yet critical enabler in managing SDM complexity.

Still, one participant highlighted an unintended consequence of integrating the chatbot into the SDM process: the introduction of a new dimension of complexity related to the potential for becoming trapped in a repetitive loop of queries and responses. While the ease of access to information was widely regarded as a benefit, it also led to a tendency to seek exhaustive data verification before advancing in the decision-making process. This behavior, driven by the desire to ensure complete and accurate information, was perceived as a potential time sink, where participants risked losing valuable time for minimal incremental gains in insight. A summary of the impact of the chatbot on dimensions of complexity can be foud in Table 2.

In conclusion, the introduction of the chatbot as a RAG-DSS in the SDM process demonstrated a tangible impact on reducing the perceived complexity of decision-making, particularly in the dimensions of "Data Extraction & Understanding" and, to a lesser extent, "Optimization & Strategic Alignment" and "Predictive Analysis & Outcome Simulation". Participants consistently emphasized the benefits of streamlined data access, structured information retrieval, and actionable insights, which collectively simplified their decision-making processes and enhanced confidence in their final decisions. However, the findings also reveal nuanced challenges, including the risk of over-reliance on iterative queries, which may inadvertently introduce inefficiencies.

5 Discussion

This study explored how integrating a RAG-enhanced DSS influences the perceived complexity of the SDM process. Our findings reveal that participants experienced a reduction in the perceived complexity of their SDM process, as well as a faster and more confident decision-making process.

The reduction in complexity observed in this study can be mapped to three distinct dimensions: Data Extraction & Understanding, Optimization & Strategic Alignment, and Predictive Analysis & Simulation. These dimensions align with the challenges commonly faced by decision-makers in complex, data-rich

Table 2. Summary of the impact of the RAG-enhanced LLM component of the DSS on the three dimensions of complexity perceived by participants.

Dimension of Complexity	Impact (+: Positive/−: Negative)
Data Extraction & Understanding	+: Easier and faster access to relevant information
	+: Information presented in a structured format
	+: Flexible structuring and granularity navigation
	+: Possibility to interpret data and make diagnoses
	−: Potential time loss due to repetitive user-chatbot interactions
Optimization & Strategic Alignment	+: Strategically sound suggestions
	+: Enhancement of user ideas with additional factors
Predictive Analysis & Outcome Simulation	+: Simulations and feedback on input strategies provided on demand

environments. The first dimension – Data Extraction & Understanding – was significantly supported by the chatbot's ability to quickly retrieve, structure, and present relevant information, allowing decision-makers to bypass the overwhelming task of manually sifting through large datasets. The second dimension – Optimization & Strategic Alignment – was addressed through the chatbot's capacity to provide strategically sound suggestions and improve the alignment of decisions with broader strategic goals. Lastly, Predictive Analysis & Simulation was facilitated by the chatbot's ability to run simulations and offer real-time feedback on potential decision outcomes.

The connection between these findings and the literature on information overload and bounded rationality is particularly noteworthy. Previous research has demonstrated that information overload is a significant barrier to effective decision-making, particularly in environments characterized by large volumes of data [4]. Our study suggests that the chatbot serves as an effective mechanism for alleviating this issue by reducing the need for exhaustive searches and enabling decision-makers to easily access only the most relevant information. By doing so, the chatbot effectively mitigates the cognitive burden typically associated with data-heavy decision-making processes. Furthermore, the chatbot appears to address some aspects of bounded rationality, allowing participants to more easily make decisions that are more informed and aligned with their goals lessening the cognitive limitations of SDM as exposed by Simon and his successors [13,16].

One notable risk is the emergence of a looping dynamic, where participants become caught in cycles of repeated queries and responses. While the chatbot facilitates faster access to information, one participant reported spending considerable time formulating follow-up questions to ensure they had gathered complete information before moving forward in the decision process. This behavior may paradoxically increase the complexity of decision-making, as individuals invest more time verifying and refining information than initially anticipated. As highlighted by Gigerenzer and Gaissmaier [7], additional information does not always lead to better decisions and may even hinder the process. This observation raises important questions about the potential paradoxical effects of RAG-DSS on decision quality. Future research should therefore explore whether a threshold exists in the utility and effectiveness of RAG-DSS, and how individual, contextual, or task-related factors may influence its efficiency in supporting decision-making.

From a practical perspective, the introduction of RAG-DSS offers several key benefits for organizations. It allows decision-makers to save time, has the potential to improve the quality of their decisions, and increases their confidence in the outcomes by providing structured, actionable insights. However, organizations should be mindful of the potential for over-reliance on the system and the risks associated with standardized decision-making. Future implementation strategies should focus on ensuring that the chatbot is efficiently used as a tool for augmentation rather than a replacement for human judgment.

Despite the promising findings, this study is subject to some limitations. First, as a case study, the generalizability of the results is constrained by the specific context of the business management simulation seminar. While this setting provided valuable insights into how participants interact with the chatbot, it may not fully capture the complexities of real-world SDM. Additionally, the limited number of participants may raise questions about the generalizability of the findings. To address these limitations, additional case studies are currently underway in organizations of varying sizes and sectors in order to test our hypotheses, enhance the validity of our results, and deepen our understanding of the mechanisms at play across diverse organizational contexts.

Future research is also planned to focus on exploring other aspects of bounded rationality, such as heuristics and biases, in the context of RAG-enhanced LLM integration into DSSs. Examining how these cognitive shortcuts are influenced by the use of RAG-DSSs could yield valuable insights into improving SDM processes. An additional factor that has yet to be examined in a future study is the potential emergence of decision isomorphism, where the chatbot's recommendations may result in standardized decisions among participants. This may reduce the diversity of decision-making strategies and lead to a convergence of strategic decisions, which could be detrimental to the innovation and risk-taking necessary for competitive differentiation in real-world environments. Finally, another important avenue for future research would be to explore the impact of RAG-DSS on the group dynamics of SDM. Strategic decisions are often made within groups [5], where various biases can emerge, such as shared information bias [19], hidden profiles [3], and groupthink [9]. In our context, some participants noted using the RAG-DSS as an arbitrator or consensus-builder. This observation suggests that the RAG-DSS could influence group decision dynamics by impacting these common decision-making biases, thus offering a potential avenue for future research into how such systems can shape the group components of SDM.

6 Conclusion

This study serves as a preliminary investigation into the potential impact of integrating RAG-enhanced LLMs into DSSs to address the inherent complexity of SDM in complex, data-rich environments. By conducting a case study in a business management simulation seminar, we sought to explore how a RAG-DSS could influence the perceived complexity of the SDM process, and if so, how.

The study involved 12 participants who interacted with a RAG-DSS during the seminar. By analyzing their chatbot interactions and conducting post-task interviews, we gained insights into both the functional role of the chatbot and its perceived impact on SDM complexity. The findings indicate that participants engaged actively with the chatbot to support various phases of the SDM process, with its impact aligning closely with the three dimensions of complexity unveiled in this study: Data Extraction & Understanding, Optimization & Strategic Resource Alignment, and Predictive Analysis & Outcome Simulation. The chatbot's ability to provide structured, contextually relevant information and actionable insights addressed the challenges of Data Extraction & Understanding, enabling participants to isolate relevant data efficiently and interpret it in a meaningful way. For Optimization & Strategic Resource Alignment, the chatbot offered strategically sound suggestions and enhancements to participants' initial ideas, supporting alignment with organizational goals. Regarding Predictive Analysis & Outcome Simulation, participants leveraged the chatbot to perform simulations and predict potential decision outcomes, although this dimension was less frequently engaged compared to the others.

Additionally, many participants described the chatbot as a reassuring tool, offering validation for their decisions and enhancing their confidence throughout the decision-making process. This multi-faceted support highlights the chatbot's potential to alleviate the complexities associated with SDM while fostering a sense of empowerment and clarity for decision-makers.

However, this study also revealed limitations and potential challenges associated with the integration of LLMs into DSSs. A participant noted the risk of falling into iterative loops of information-seeking interactions with the chatbot, which could inadvertently increase decision time. Additionally, the potential for decision isomorphism, where reliance on chatbot-generated suggestions might lead to homogenized decision-making across users, could limit creativity and strategic differentiation. These limitations highlight important areas for future research, including investigating strategies to mitigate iterative interaction loops, exploring mechanisms to preserve creativity and strategic uniqueness, and assessing the broader implications of decision isomorphism in real-world organizational settings.

In conclusion, this study highlights the significant potential of RAG-DSSs to reduce the complexity of SDM and boost decision-maker confidence in navigating data-intensive environments. By addressing both the opportunities and challenges associated with this transformative technology, future research can deepen our understanding of its impact and refine its integration to better support and enhance strategic decision-making processes.

References

1. Bhatt, G.D., Zaveri, J.: The enabling role of decision support systems in organizational learning. Decis. Support Syst. **32**(3), 297–309 (2002)
2. Bonczek, R.H., Holsapple, C.W., Whinston, A.B.: Foundations of Decision Support Systems. Academic Press (2014)
3. Dayeh, V., Morrison, B.W.: The effect of perceived competence and competitive environment on team decision-making in the hidden-profile paradigm. Group Decis. Negot. **29**(6), 1181–1205 (2020)
4. Edmunds, A., Morris, A.: The problem of information overload in business organisations: a review of the literature. Int. J. Inf. Manage. **20**(1), 17–28 (2000)
5. Eisenhardt, K.M., Zbaracki, M.J.: Strategic decision making. Strateg. Manag. J. **13**(S2), 17–37 (1992)
6. Feuerriegel, S., Hartmann, J., Janiesch, C., Zschech, P.: Generative AI. Bus. Inf. Syst. Eng. **66**(1), 111–126 (2024)
7. Gigerenzer, G., Gaissmaier, W.: Heuristic decision making. Annu. Rev. Psychol. **62**(2011), 451–482 (2011)

8. Gioia, D.A., Corley, K.G., Hamilton, A.L.: Seeking qualitative rigor in inductive research: Notes on the Gioia methodology. Organ. Res. Methods **16**(1), 15–31 (2013)
9. Janis, I.L.: Groupthink. IEEE Eng. Manage. Rev. **36**(1), 36 (2008)
10. Josselson, R., Hammack, P.L.: Essentials of Narrative Analysis. American Psychological Association (2021)
11. Lewis, P., et al.: Retrieval-augmented generation for knowledge-intensive NLP tasks. Adv. Neural. Inf. Process. Syst. **33**, 9459–9474 (2020)
12. Melnyk, S.A., Narasimhan, R., DeCampos, H.A.: Supply chain design: Issues, challenges, frameworks and solutions. Int. J. Prod. Res. **52**(7), 1887–1896 (2014)
13. Mintzberg, H., Raisinghani, D., Theoret, A.: The structure of "unstructured" decision processes. Adm. Sci. Q. 246–275 (1976)
14. Nutt, P.C.: Types of organizational decision processes. Adm. Sci. Q. 414–450 (1984)
15. Power, D.J.: Understanding data-driven decision support systems. Inf. Syst. Manag. **25**(2), 149–154 (2008)
16. Simon, H.A.: The Shape of Automation. Harper and Row, New York (1965)
17. Teubner, T., Flath, C.M., Weinhardt, C., van der Aalst, W., Hinz, O.: Welcome to the era of chatGPT et al. the prospects of large language models. Bus. Inf. Syst. Eng. **65**(2), 95–101 (2023)
18. Trunk, A., Birkel, H., Hartmann, E.: On the current state of combining human and artificial intelligence for strategic organizational decision making. Bus. Res. **13**(3), 875–919 (2020). https://doi.org/10.1007/s40685-020-00133-x
19. Van Swol, L.M.: Perceived importance of information: the effects of mentioning information, shared information bias, ownership bias, reiteration, and confirmation bias. Group Process. Intergroup Relat. **10**(2), 239–256 (2007)
20. Yin, R.K.: Case Study Research: Design and Methods, 6th edn. Sage, Thousand Oaks (2017)

Cross-Lingual Entity Linking Using GPT Models in Radiology Abstracts

Mariana Dias(✉) and Carla Teixeira Lopes

INESC TEC, Faculty of Engineering, University of Porto, Porto, Portugal
{up201606486,ctl}@fe.up.pt

Abstract. Entity linking is an important task in medical natural language processing (NLP) for converting unstructured text into structured data for clinical analysis and semantic interoperability. However, in lower-resource languages, this task is challenging due to the limited availability of domain-specific resources. This paper explores a translation-based cross-lingual entity linking approach using GPT models, GPT-3.5 and GPT-4o, for zero-shot machine translation and entity linking with in-context learning. We evaluate our approach using a Portuguese-English parallel dataset of radiology abstracts. Our results show that chunk-level machine translation outperforms sentence-level translation. Moreover, our translation-based approach to cross-lingual entity linking of UMLS concepts outperformed the multilingual encoder method baseline. However, the in-context learning entity linking approach did not outperform a translation-based approach with a dictionary-based entity linking method.

Keywords: Medical Entity Linking · Large Language Models · GPT

1 Introduction

Entity linking is an important task in medical natural language processing (NLP), especially in clinical settings where large volumes of unstructured text require analysis and interpretation. By linking entity mentions in these documents to standardized concepts in medical terminologies, we can transform unstructured textual documents into a structured format more suitable for clinical analysis and decision support and ensure semantic interoperability [16].

Most medical ontologies and vocabularies are primarily available in English, limiting their use in lower-resource languages for various NLP tasks, including entity linking. Current state-of-the-art approaches use transformer models for cross-lingual entity linking. These models leverage multilingual encoders to align entity mentions across languages, often requiring pre-training and fine-tuning on large-scale data.

Recent advances in Large Language Models (LLM), particularly generative pre-trained transformer (GPT) models, have unlocked new possibilities to solve NLP tasks that do not require in-domain or task-specific training [4]. To our knowledge, no prior research has explored cross-lingual entity linking using GPT models in a translation-based framework. This paper aims to analyze the potential of GPT models for cross-lingual entity linking through machine translation, and in-context learning for entity recognition, alignment with an ontology, and projection. In this context, alignment refers to associating entity mentions in the translated text with standardized entities from an ontology, and projection involves transferring the linked entities back to the original language. Our experiment focuses on a radiology dataset [3], linking entity mentions to a standardized radiology ontology, RadLex.

We aim to address the following research questions:

1. How does the granularity of prompt context impact GPT models' performance in machine translation of radiology-related data?
2. Do larger, more advanced GPT models achieve better results than smaller ones in the entity linking task?
3. Do GPT models outperform other approaches for entity linking?

2 Cross-Lingual Entity Linking

Given a textual document D in a source language L_S, the goal of the cross-lingual entity linking task is to identify entity mentions m_1, ..., m_n within D and link each m_i mention to an entity $E_j \in KB$, where KB is a knowledge base in the target language L_T containing a set of entities $\{E_1, ..., E_m\}$, like the Unified Medical Language System (UMLS). The UMLS metathesaurus [1] is a well-known biomedical knowledge base that integrates various vocabularies.

Most works formulate the entity linking task as a multi-class classification or ranking problem [18]. Recent approaches use transformer-based models to build dense entity representations and compute relevance scores for entity candidates. Botha et al. [2] developed a bi-encoder model with mention and entity encoders initialized from pre-trained multilingual BERT models. Their method embeds mention-entity pairs in a shared vector space to retrieve entity candidates. Their approach outperformed others on the TR2016^{hard} dataset, including Upadhyay et al.'s [19] FastText-based method.

In the biomedical domain, Liu et al. [10] developed cross-lingual variations of SapBERT [9], a biomedical BERT-based model fine-tuned on UMLS synonyms. Their approach, leveraging multilingual encoders MBERT and XLMR, outperformed monolingual models on the cross-lingual biomedical entity linking benchmark (XL-BEL)[1] in lower-resource languages linguistically distant from Romance and Germanic languages.

Recent research has explored the use of GPT models for entity linking, particularly through in-context learning. Shlyk et al. [17] created a retrieval-augmented entity linking approach for biomedical concepts using in-context learning prompts. Groza et al. [6] evaluated GPT models for linking phenotype concepts through in-context learning. Both studies reported competitive performance on benchmark datasets. Other approaches, such as Ding et al.'s [5], leverage prompt engineering and instruction tuning to improve entity linking performance.

---

[1] https://paperswithcode.com/dataset/xl-bel

3 Methodology

The cross-lingual entity linking pipeline consists of three phases: (1) translating a document D in a radiology dataset from the source language L_S (Portuguese) to the target language L_T (English), (2) recognizing entity mentions m_i and aligning them to terms E in the RadLex ontology, and (3) back-translating the annotated document D_a to L_S (Portuguese). Figure 1 provides an overview of the system's architecture with examples of outputs from each phase.

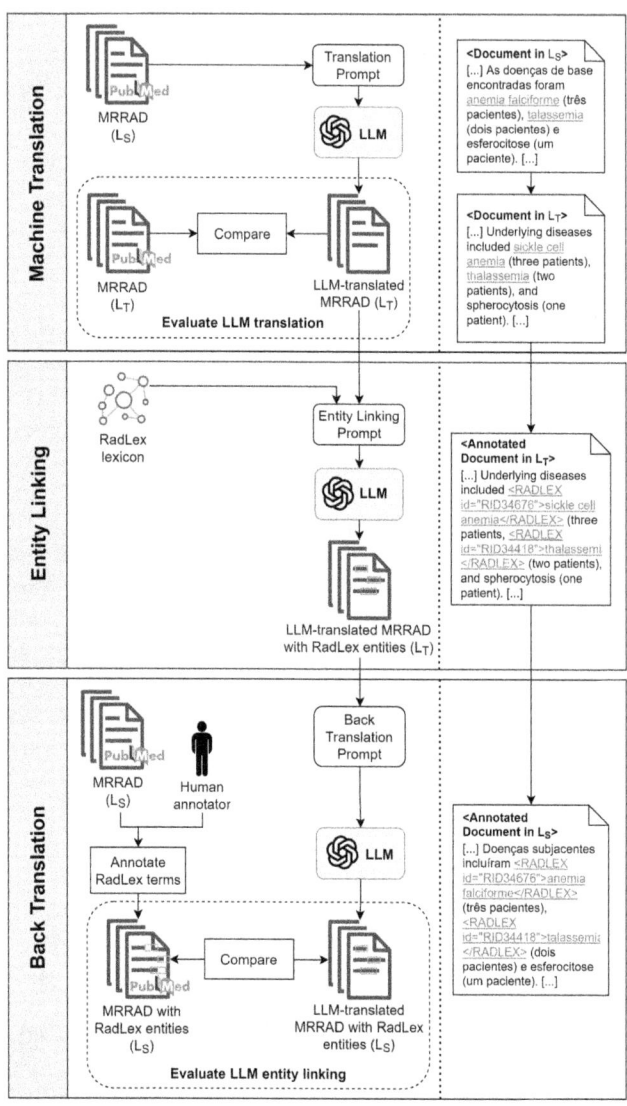

Fig. 1. Architecture of an entity linking system for RadLex entities in the MRRAD dataset.

Our pipeline employs a domain-specific ontology and parallel corpus for evaluation. The knowledge base used is the RadLex[2] lexicon, developed by the Radiological Society of North America (RSNA). The RadLex ontology consists of 46,761 classes, of which 1,323 are linked to UMLS concepts. We evaluated our approach with the Multilingual Radiology Research Articles Dataset[3] (MRRAD) [3], a Portuguese-English parallel corpus that contains 34 PubMed abstracts related to radiology. Table 1 summarizes the dataset statistics.

Table 1. MRRAD dataset statistics: number of documents, average number of sentences per document, and average number of words per document.

Language	# Documents	Avg. # Sentences/Doc	Avg. # Words/Doc
Portuguese	34	123.6	2,947.2
English	34	151.7	2,908.4

Our goal with this study is to assess the feasibility of a three-stage LLM-based translation approach for cross-lingual entity linking. To achieve this, we compared the performance of two proprietary models from OpenAI, GPT-3.5 and GPT-4o, using zero-shot machine translation, in-context learning entity linking with pre-filtered ontology terms, and different prompting strategies. As a baseline, we included a system that combines GPT-based machine translation and back translation with dictionary lookup for entity linking and projection.

3.1 Machine Translation

For machine translation, we proposed two task-specific prompts with different granularities: sentence-level and word-chunk fitted to the LLM's context window. In both prompts, past queries and responses are retained to maintain context. The sentence-level prompt uses a full sentence as input, while the chunk-level prompt uses word chunks obtained by tokenizing the text with OpenAI's tiktoken[4] tokenizer and splitting it based on the LLM's context window as the threshold for maximum chunk size. Moreover, the chunk-level prompt uses a format that differentiates between the first and the subsequent chunks. We present the machine translation prompting approaches in Fig. 2.

---

[2] https://www.rsna.org/radlex/.
[3] https://github.com/lasigeBioTM/MRRAD.
[4] https://github.com/openai/tiktoken.

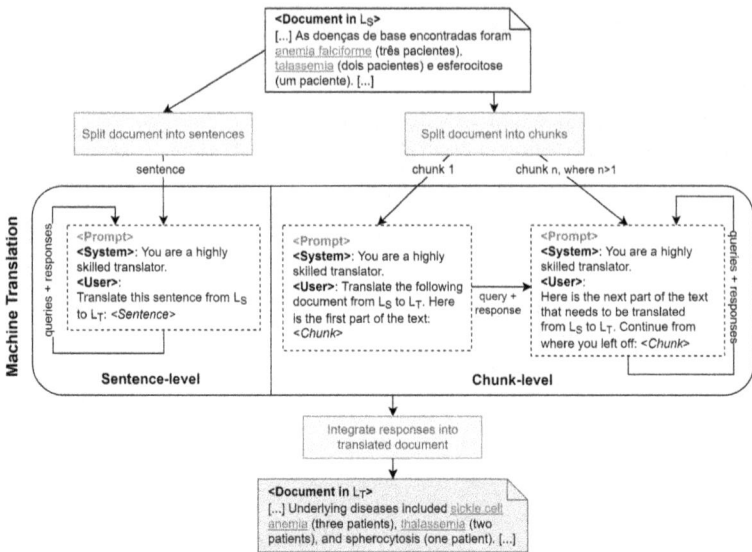

Fig. 2. Machine translation prompt experiment with example.

3.2 Entity Linking

Following the work of Hu et al. [7] of prompt engineering for clinical named entity recognition, we designed two prompting approaches for entity linking: an original prompt with a task description, format specification, and context, and a subsequently refined prompt. Figure 3 illustrates the entity linking prompting process, including the used prompts.

The original prompt follows a structured format that includes a task description, a format specification, and context to guide entity recognition and alignment with RadLex terms. We instruct models to use HTML tags to annotate entity mentions and their linked RadLex entities. The input consists of a sentence from a translated radiology abstract, supplemented with a list of relevant RadLex terms and their identifiers. We generate a list of candidate RadLex terms for each sentence using a dictionary lookup approach. We identify relevant RadLex terms and synonyms while filtering out shorter terms, retaining only those longer than three characters.

The subsequent refined prompt provides more detailed formatting instructions based on an analysis of the results from the initial prompt. It instructs the models to use valid tag syntax and identifiers by addressing common errors identified with the initial prompt. We formulated and refined five rules using ChatGPT: 1) use only provided term-id pairs to reduce hallucinations of non-existent RadLex terms or identifiers, 2) enforce identifier formatting, 3) prohibit entity names as identifiers, 4) ensure proper tag syntax to mitigate improperly closed tags, and 5) instruct the model to return the original sentence if no entities are recognized.

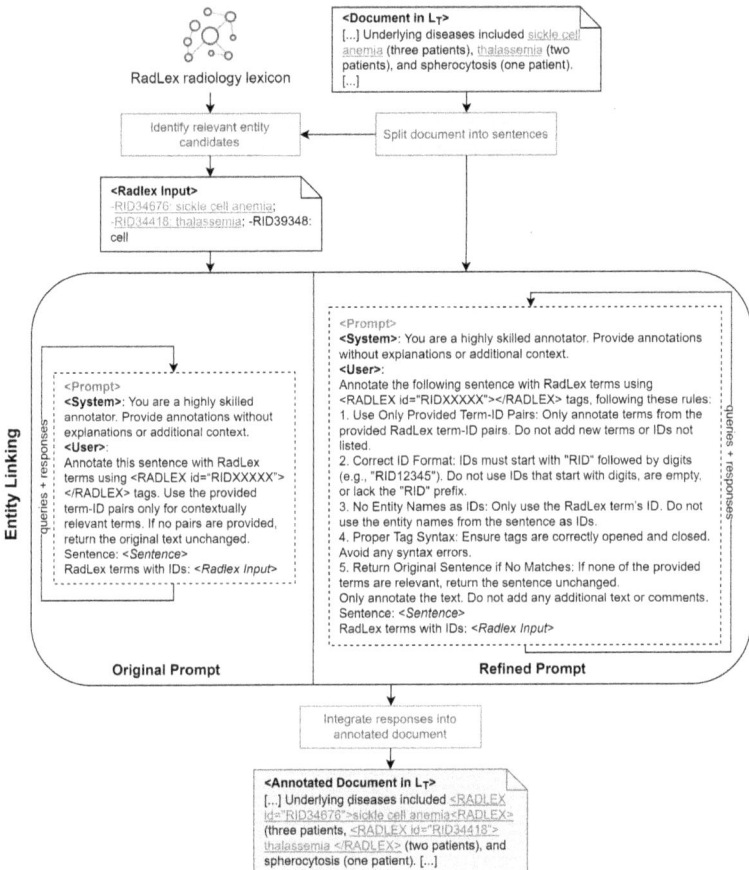

Fig. 3. Entity linking prompting experiment with example.

To evaluate the entity linking phase, we use two approaches as baselines: a dictionary-based approach through the NCBO annotator and a multilingual encoder-based method. The NCBO annotator [8], developed by the National Center for Biomedical Ontology (NCBO), annotates biomedical documents by matching terms to a dictionary built from ontologies hosted in BioPortal[5]. We integrated this approach into our pipeline by replacing the GPT-based entity linking stage with the NCBO Annotator while maintaining the machine translation and back translation steps. We accessed the BioPortal REST API[6] through the Annotator endpoint with default parameter settings[7]. For the multilingual

---

[5] https://bioportal.bioontology.org/ontologies.
[6] https://data.bioontology.org/annotator.
[7] For more information, consult the documentation at https://data.bioontology.org/documentation.

encoder baseline, we used SapBERT-UMLS-2020AB-all-lang-from-XLMR[8] [10], a SapBERT model trained on UMLS, to generate dense embeddings to represent RadLex entities. We generated candidate entity mentions using an n-gram approach and performed entity linking by computing similarity scores between mention embeddings and RadLex entity embeddings. We linked mentions to RadLex terms when the similarity score exceeded a threshold of 0.9. Figure 4 demonstrates the pipeline for the baseline approaches.

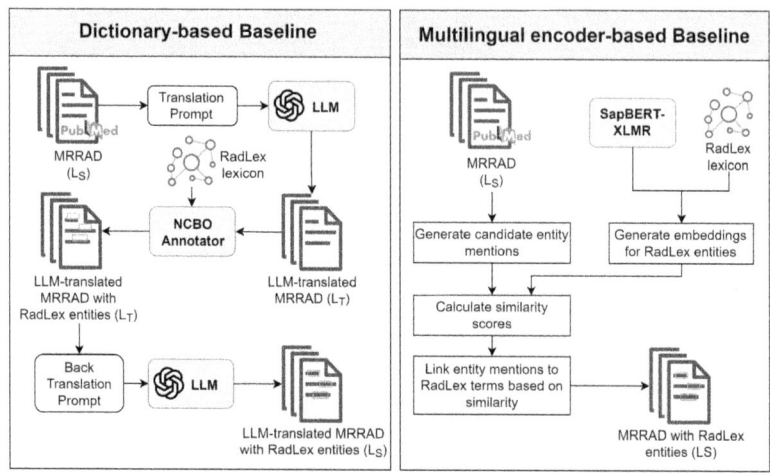

Fig. 4. Cross-lingual entity linking baseline pipeline.

3.3 Back Translation

We performed back translation on the annotated text containing RadLex entities, projecting the tags from the target language L_T to the source language L_S through the translation process. To maintain the integrity of the HTML-like tags that identify entity mentions, we did not use a chunk-based approach, as used in the machine translation phase, to prevent cutting off entities. Instead, we used a sentence-level back translation prompt with an additional instruction to preserve HTML tags in the output to ensure that the structure of the original text is maintained while incorporating the linked entities. Initially, we used the same prompt as for machine translation and refined it based on results from experiments on a few documents. For the final prompt, we consulted ChatGPT for suggestions on potential prompts that could decrease errors. Figure 5 shows the back translation prompting process.

---

[8] https://huggingface.co/cambridgeltl/SapBERT-UMLS-2020AB-all-lang-from-XLMR.

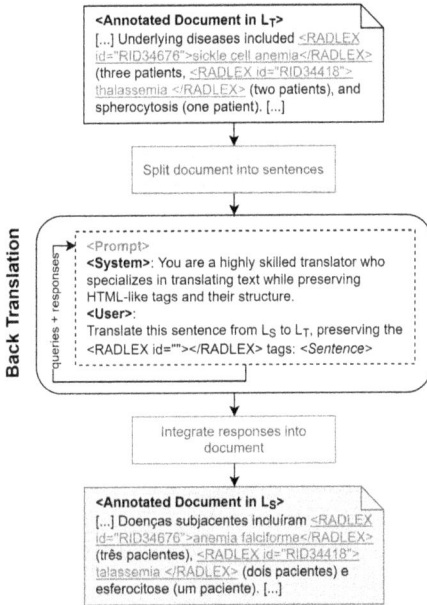

Fig. 5. Back translation prompting experiment with example.

To evaluate the performance of our approach in cross-lingual entity linking, we manually created a gold standard for the MRRAD dataset by annotating each document with RadLex entity mentions. We performed the annotation using the Protégé ontology editor with the Knowtator plugin[9]. During the annotation process, we focused on ontology classes that represent biomedical terms to ensure linking to relevant radiological concepts in diverse medical contexts. The result-

Table 2. Classes selected for gold standard annotation with total of descendent nested subclasses.

Class	Name	# Descendant Classes
RID3	Anatomical Entity	38,165
RID34785	Clinical Finding	2,230
RID5	Imaging Observation	1,133
RID50606	Imaging Specialty	86
RID7479	Non-anatomical Substance	392
RID34861	Object	403
RID1559	Procedure	610
RID39128	Process	35

---

[9] https://github.com/UCDenver-ccp/Knowtator-2.0.

ing gold standard dataset contains 4,327 linked entities, averaging 127 linked entities per document. Table 2 presents the selected classes and the number of nested subclasses for annotation.

4 Results

We divided the experiments into three phases: document-level machine translation quality evaluation in Sect. 4.1, entity linking error analysis in Sect. 4.2, and cross-lingual entity linking evaluation in Sect. 4.3.

4.1 Document-Level Machine Translation

We report the results of executing the machine translation prompts described in Sect. 3.1 in Table 3. To evaluate the translation quality, we used measures that assess lexical precision, BLEU [11] and ChrF++ [12] with SacreBLEU[10] [13], and neural metrics that evaluate semantic accuracy, COMETkiwi [15] (wmt22-COMETkiwi-da[11]) and COMET-22 [14] (wmt22-COMET-da[12]).

Table 3. Machine translation performance of GPT models on MRRAD dataset.

System	BLEU	ChrF++	COMETkiwi	COMET-22
Prompt S				
GPT-3.5	50.85	**88.29**	58.73	88.11
GPT-4o	**52.27**	81.06	60.86	88.42
Prompt C				
GPT-3.5	36.50	66.40	61.68	88.48
GPT-4o	33.81	64.43	**62.34**	**88.50**

The chunk-level prompt (Prompt C) performs better than the sentence-level prompt (Prompt S) with neural-based COMET measures. However, it performs worse using lexical-based measures like BLEU and ChrF++. We performed a qualitative analysis of machine translation outputs generated using different prompts and GPT models to better understand why the lexical-based measures declined in performance with GPT-4o, while the neural-based metrics improved. We present examples of translations that illustrate that lexical-based measures are likely more sensitive to exact word matches and less adaptable to variations in vocabulary and phrasing than neural-based metrics. Listing 1.1 demonstrates an example of machine translation outputs with the sentence-level prompt where the GPT-4o translation had more variations in vocabulary and phrasing.

---

[10] https://github.com/mjpost/sacrebleu.
[11] https://huggingface.co/Unbabel/wmt22-cometkiwi-da.
[12] https://huggingface.co/Unbabel/wmt22-comet-da.

Listing 1.1: Machine translation outputs using Prompt S.

Original sentence:
RESUMO OBJETIVO: Descrever a distribuição dos escores de cálcio coronariano numa população de homens brasileiros brancos assintomáticos submetidos à avaliação pela tomografia ultra-rápida.
Reference translation:
ABSTRACT OBJETIVE: To describe the distribution of coronary artery calcium scores in a population of asymptomatic white Brazilian men undergoing assessment with ultrafast computed tomography.
Translation with GPT-3.5:
SUMMARY OBJECTIVE: To describe the distribution of coronary calcium scores in a population of asymptomatic Brazilian white men undergoing evaluation by ultrafast computed tomography.
Translation with GPT-4o:
OBJECTIVE SUMMARY: Describe the distribution of coronary calcium scores in a population of asymptomatic white Brazilian men assessed using ultra-fast tomography.

Listing 1.2 shows an example of machine translation outputs with the chunk-level prompt where GPT-4o preserves the full citation of the PEPI program and maintains the original sentence flow, whereas GPT-3.5 omits the citation and splits the sentence into two.

Listing 1.2: Machine translation outputs using Prompt C.

Original sentence:
Os pacientes foram divididos randomicamente utilizando o programa PEPI {[COMPUTER PROGRAMS FOR EPIDEMIOLOGISTS (PEPI)] by J.H. Abramson and Paul M. Gahlinger. Version 4.04x}, em dois grupos com sorteio de envelope selado, as seringas contendo a droga eram preparadas por um pesquisador que não fosse avaliar o paciente.
Reference translation:
The patients were randomly divided into two groups using a raffle with sealed envelops and the computer program PEPI (Computer Programs for Epidemiologists by J.H. Abramson and Paul M. Gahlinger. Version 4.04x). The syringes containing the drug were prepared by a researcher that would not evaluate the patients.
Translation with GPT-3.5:
The patients were randomly divided into two groups using the PEPI program, with sealed envelope randomization. The syringes containing the drug were prepared by a researcher who did not evaluate the patient.
Translation with GPT-4o:
The patients were randomly divided using the PEPI program (COMPUTER PROGRAMS FOR EPIDEMIOLOGISTS (PEPI) by J.H. Abramson and Paul M. Gahlinger. Version 4.04x) into two groups with sealed envelope allocation, and the syringes containing the drug were prepared by a researcher who would not evaluate the patient.

Based on the previous analysis, we have decided to prioritize neural-based measures. Thus, we conducted statistical tests to evaluate the machine translation quality difference between the two prompts and the two GPT models using the COMET-based metrics. As the data was paired, we initially considered conducting paired t-tests. However, upon assessing normality and outliers assumptions, we found that the reference-based COMET-22 and reference-free COMETkiwi metrics did not meet normal distribution requirements. Therefore, we used the Wilcoxon signed-rank test as a non-parametric alternative to the paired t-test. We used this test to compare 1) the mean difference between the two prompts for each model, and 2) the difference between the two models using the same prompts. Table 4 presents the Wilcoxon signed-rank test results.

Table 4. p-values of Wilcoxon signed-rank test pairwise comparisons.

Comparison	COMETkiwi	COMET-22
GPT-3.5$_{PromptS}$ < GPT-3.5$_{PromptC}$	<.001	<.01
GPT-4o$_{PromptS}$ < GPT-4o$_{PromptC}$	<.01	.289
GPT-3.5$_{PromptS}$ < GPT-4o$_{PromptS}$	<.001	<.01
GPT-3.5$_{PromptC}$ < GPT-4o$_{PromptC}$	.163	.361

For both COMETkiwi and COMET-22, GPT-4o significantly outperforms GPT-3.5, indicating an advantage of GPT-4o over GPT-3.5 when using Prompt S. However, there are no significant differences for either metric with Prompt C. For the comparison between prompts with the same models, there are significant differences for both metrics with GPT-3.5. This suggests that, for the GPT 3.5 model, Prompt C produces higher COMETkiwi and COMET-22 scores than Prompt S. Since Prompt C demonstrates superior performance in machine translation, we will use the documents translated with this prompt for the entity linking task.

4.2 Entity Linking Error Analysis

To assess the performance of our prompting strategies in the entity linking task, we analyzed errors in the entity linking and back translation phases, focusing on hallucinations and their impact on entity linking performance. In this context, we consider hallucinations as invalid RadLex identifiers generated by the GPT models. To understand the nature of the hallucinations, we analyzed and categorized the misrepresented RadLex identifiers that caused them. We classified common linking errors into five types: missing, no prefix, numeric, invalid, and textual. Table 5 provides definitions and examples for each error type.

Table 5. Entity linking error typology with examples.

Error type	Description	Example
Missing	Entity mentions without a RadLex identifier	"Os registros médicos de 3.101 <RADLEX id=""></RADLEX> vítimas [...]"
No prefix	Numeric identifiers that are RadLex terms but lack the "RID" prefix	"[...] quantificação do <RADLEX id="11800">cálcio</RADLEX>"
Numeric	Numeric identifiers that are not RadLex terms	"Considerado significativo quando alpha <RADLEX id="12345">0,05</RADLEX>"
Textual	Entity mentions with textual identifier	"Esta <RADLEX id="disorder">condição</RADLEX> é rara [...]"
Invalid	Identifiers that follows a valid format but are not RadLex terms	"[...] aspectos clínicos e radiográficos <RADLEX id="RID12940">correspondentes</RADLEX>"

We compared the performance of the dictionary-based baseline and GPT models using different entity linking prompts across both phases. Table 6 shows the frequency of RadLex entity mentions identified and associated hallucination rates.

Regarding the baseline, using the NCBO Annotator for entity linking and the GPT models for machine and back translation resulted in a higher hallucination rate with GPT-3.5, suggesting that GPT-4o is slightly more reliable in generating accurate RadLex identifiers. The GPT models exhibited lower hallucination rates in the entity linking stage compared to the back translation phase, with GPT-4o achieving the lowest hallucination rates near 0%. In the back translation stage, GPT-4o still maintained low hallucination rates of 1.22%-1.42%. The GPT-3.5 model identified more total and unique terms with the refined entity linking prompt than with the initial prompt and had a slightly lower hallucination rate. We also observed a reduction in hallucination rates with the refined prompt compared to the original prompt in all observations.

We analyzed the distribution of the classified RadLex identifier errors across different approaches, including the dictionary-based baseline, and GPT models with different prompts in the entity linking and back translation phases, as illustrated in Fig. 6.

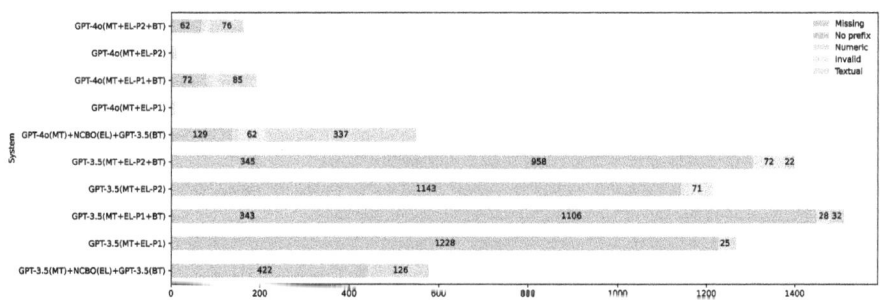

Fig. 6. Distribution of RadLex identifier errors.

Consistent with the earlier analysis, where entity linking approaches exhibited lower hallucination rates, the methods in the entity linking phase had significantly fewer errors, particularly missing and textual errors. This suggests that these error types were likely introduced during the back translation process.

The NCBO baseline with the GPT-4o model generally produced fewer missing errors compared to the baseline with GPT-3.5. However, it showed a higher occurrence of invalid and textual errors. The refined entity linking prompt seems to have greatly decreased the occurrence of no prefix errors in the two GPT models, although it led to an increase in invalid errors with the GPT-3.5 model. Overall, the GPT-4o model outperformed the GPT-3.5 model in minimizing missing, no prefix, and invalid errors but exhibits a slightly higher frequency of textual errors.

Table 6. Overview of frequency of RadLex terms identified across all experiments and hallucination rates.

System	Total	Unique	Hallucination Rate (%)
Ground truth	4,327	927	-
GPT-3.5$_{MT}$+NCBO$_{EL}$+GPT-3.5$_{BT}$	6,641	3,316	8.70
GPT-4o$_{MT}$+NCBO$_{EL}$+GPT-4o$_{BT}$	7,136	3,713	7.71
GPT-3.5$_{MT+EL-P1}$	13,824	5,161	9.17
GPT-3.5$_{MT+EL-P2}$	14,235	5,215	8.53
GPT-4o$_{MT+EL-P1}$	13,032	3,984	**0.06**
GPT-4o$_{MT+EL-P2}$	13,036	3,999	0.08
GPT-3.5$_{MT+EL-P1+BT}$	12,806	4,876	11.81
GPT-3.5$_{MT+EL-P2+BT}$	13,207	4,907	10.59
GPT-4o$_{MT+EL-P1+BT}$	13,523	4,105	1.42
GPT-4o$_{MT+EL-P2+BT}$	13,336	4,105	1.22

MT: machine translation, EL-P1: original entity linking prompt, EL-P2: refined entity linking prompt, BT: back translation.

We proceeded to perform a qualitative analysis of textual errors generated by the GPT-3.5 and GPT-4o models to understand if there is a correlation between the back translation prompt and the induction of hallucinations. In Listing 1.3, we provide an example of an output that demonstrates how all observed textual errors originated from sentences that did not contain RadLex entity mentions during the entity linking step, but were incorrectly annotated with the back translation prompt.

Listing 1.3: Entity linking output with textual error during back translation stage.

Original sentence:
Recentemente, desenvolvemos um sistema de visão computacional, o qual denominamos SIStema para a Detecção e a quantificação de Enfisema Pulmonar (SISDEP).
Machine Translation output:
We recently developed a computer vision system, named Pulmonary Emphysema Detection and Quantification System (SISDEP).
Entity Linking output:
We recently developed a computer vision system, named Pulmonary Emphysema Detection and Quantification System (SISDEP).
Back Translation output:
Desenvolvemos recentemente um sistema de visão computacional, chamado <RADLEX id="Pulmonary_Emphysema">Sistema de Detecção e Quantificação de Enfisema Pulmonar</RADLEX> (SISDEP).

4.3 Cross-Lingual Entity Linking Evaluation

To assess the effectiveness of our approach in biomedical cross-lingual entity linking, we focused on a subset of the RadLex ontology that contains standardized UMLS concepts. We used document-level precision (P), recall (R), and F1-score (F1) metrics, which are calculated based on the overlap between predicted entity mentions and their corresponding RadLex terms in each document. To evaluate the impact of hallucinations, we perform post-processing on the back translation step to filter out invalid RadLex entity links. Table 7 presents the document-level evaluation results, comparing the performance of our approach to the dictionary-based and multilingual encoder baselines.

The translation-based approach to entity linking outperformed the multilingual encoder-based SapBERT-XLMR model. Our approach performed the best with the dictionary-based NCBO annotator for entity linking compared to the in-context learning approach of the GPT models. For the dictionary-based baseline, the choice of GPT model had a minimal impact on performance.

Applying post-processing to filter out hallucinations enables a more accurate assessment of the approaches' performance, as any observed decrease in precision is more likely to reflect the approach's limitations in identifying relevant entities, rather than errors caused by hallucinated identifiers. The entity linking prompt strategies had no performance impact with the GPT-4o model. In contrast, for the GPT-3.5 model, the refined entity linking prompt resulted in a slight improvement in recall, with no effect on precision, suggesting that the refined prompt was more effective at identifying relevant UMLS concepts from the RadLex ontology.

Table 7. Document-level entity linking evaluation results with UMLS terms.

System	P	R	F1
SapBERT-XLMR$_{base}$	0.15	0.17	0.14
GPT-3.5$_{MT}$			
+ NCBO$_{EL}$+GPT-3.5$_{BT-NoPP}$	0.29	**0.78**	0.40
+ GPT-3.5$_{EL-P1+BT-NoPP}$	0.15	0.73	0.24
+ GPT-3.5$_{EL-P2+BT-NoPP}$	0.17	0.76	0.26
+ NCBO$_{EL}$+GPT-3.5$_{BT-WithPP}$	**0.38**	**0.78**	**0.49**
+ GPT-3.5$_{EL-P1+BT-WithPP}$	0.32	0.73	0.42
+ GPT-3.5$_{EL-P2+BT-WithPP}$	0.31	0.76	0.42
GPT-4o$_{MT}$			
+ NCBO$_{EL}$+GPT-4o$_{BT-NoPP}$	0.24	**0.78**	0.34
+ GPT-4o$_{EL-P1+BT-NoPP}$	0.28	0.76	0.38
+ GPT-4o$_{EL-P2+BT-NoPP}$	0.28	**0.78**	0.39
+ NCBO$_{EL}$+GPT-4o$_{BT-WithPP}$	0.37	0.78	0.48
+ GPT-4o$_{EL-P1+BT-WithPP}$	0.33	0.76	0.43
+ GPT-4o$_{EL-P2+BT-WithPP}$	0.33	0.78	0.44

MT: machine translation, BT: back translation, EL-P1: original entity linking prompt, EL-P2: refined entity linking prompt, NoPP: without post-processing, WithPP: with post-processing.

5 Discussion

The machine translation evaluation demonstrated that prompt choice had a significant impact on GPT-3.5's performance, whereas GPT-4o was less influenced by prompt variations. This finding addresses RQ1, confirming that the granularity of a prompt's context impacts model performance. We also concluded that GPT-4o significantly outperformed GPT-3.5, indicating that, in the machine translation phase, a larger model achieved the best results, addressing RQ2.

In the cross-lingual entity evaluation, we did not find any major differences in performance between GPT-3.5 and GPT-4o or between the different entity linking prompt strategies. This suggests that, relating to RQ2, increasing model size did not lead to improvements in entity linking performance unlike in machine translation.

Regarding our translation-based entity linking approach, the dictionary-based approach achieved comparable or superior F1-scores in comparison to the in-context learning method with GPT models. In response to RQ3, GPT models did not outperform the dictionary-based approaches for entity linking, indicating that GPT models did not significantly enhance the contextual recognition of relevant entities.

6 Conclusion

In this study, we explored the use of GPT models, specifically GPT-3.5 and GPT-4o, in the cross-lingual entity linking task using a translation-based approach that consists of three phases: machine translation, entity linking, and back translation. We explored different prompting strategies and entity linking approaches, including a dictionary-based method and in-context learning.

In the machine translation phase, our results showed that chunk-level machine translation outperformed sentence-level translation in the MRRAD dataset. During the entity linking phase, our error analysis revealed that the GPT-4o model had a near 0% hallucination rate. In the back translation phase, when evaluating cross-lingual entity linking with UMLS terms in the RadLex ontology, our approach outperformed the baseline multilingual encoder-based method. However, the in-context learning entity linking approach did not outperform the dictionary-based method.

Overall, our translation-based approach to cross-lingual entity linking shows potential as a viable method, but its effectiveness should be further evaluated on a wider range of datasets to assess its robustness. While post-processing helped mitigate hallucinations, it could not overcome the limitations of GPT models in accurately linking entities.

For future work, it would be interesting to explore other LLMs beyond GPT models and implement a knowledge retriever for ambiguous entities that could further enrich the prompt context and improve model performance.

Acknowledgments. This work is co-financed by Component 5 - Capitalization and Business Innovation, integrated in the Resilience Dimension of the Recovery and Resilience Plan within the scope of the Recovery and Resilience Mechanism (MRR) of the European Union (EU), framed in the Next Generation EU, for the period 2021–2026, within project HfPT, with reference 41.

References

1. Bodenreider, O.: The unified medical language system (UMLS): integrating biomedical terminology. Nucleic Acids Res. **32** (2004). https://doi.org/10.1093/nar/gkh061
2. Botha, J.A., Shan, Z., Gillick, D.: Entity linking in 100 languages. In: EMNLP 2020 - 2020 Conference on Empirical Methods in Natural Language Processing, Proceedings of the Conference (2020). https://doi.org/10.18653/v1/2020.emnlp-main.630
3. Campos, L., Pedro, V., Couto, F.: Impact of translation on named-entity recognition in radiology texts. Database: J. Biol. Databases Curation **2017** (2017). https://doi.org/10.1093/database/bax064
4. Ding, B., et al.: Is GPT-3 a good data annotator? In: Proceedings of the Annual Meeting of the Association for Computational Linguistics, vol. 1 (2023). https://doi.org/10.18653/v1/2023.acl-long.626

5. Ding, Y., Poudel, A., Zeng, Q., Weninger, T., Veeramani, B., Bhattacharya, S.: EntGPT: linking generative large language models with knowledge bases (2024). https://doi.org/10.48550/arXiv.2402.06738
6. Groza, T., et al.: An evaluation of GPT models for phenotype concept recognition. BMC Med. Inform. Decis. Making **24** (2024). https://doi.org/10.1186/s12911-024-02439-w
7. Hu, Y., et al.: Improving large language models for clinical named entity recognition via prompt engineering. J. Am. Med. Inform. Assoc. **31** (2024). https://doi.org/10.1093/jamia/ocad259
8. Jonquet, C., Shah, N.H., Youn, C.H., Musen, M.A., Callendar, C., Storey, M.A.: NCBO annotator: semantic annotation of biomedical data. In: 8th International Semantic Web Conference, Poster and Demo Session (ISWC 2009), Washington, DC, USA (2009). https://hal.science/hal-04276274
9. Liu, F., Shareghi, E., Meng, Z., Basaldella, M., Collier, N.: Self-alignment pretraining for biomedical entity representations. In: NAACL-HLT 2021 - 2021 Conference of the North American Chapter of the Association for Computational Linguistics: Human Language Technologies, Proceedings of the Conference (2021). https://doi.org/10.18653/v1/2021.naacl-main.334
10. Liu, F., Vulić, I., Korhonen, A., Collier, N.: Learning domain-specialised representations for cross-lingual biomedical entity linking. In: ACL-IJCNLP 2021 - 59th Annual Meeting of the Association for Computational Linguistics and the 11th International Joint Conference on Natural Language Processing, Proceedings of the Conference, vol. 2 (2021). https://doi.org/10.18653/v1/2021.acl-short.72
11. Papineni, K., Roukos, S., Ward, T., Zhu, W.J.: BLEU: a method for automatic evaluation of machine translation. In: Proceedings of the 40th Annual Meeting on Association for Computational Linguistics, ACL 2002, pp. 311–318. Association for Computational Linguistics, USA (2002). https://doi.org/10.3115/1073083.1073135
12. Popovic, M.: CHRF ++: words helping character n-grams. In: WMT 2017 - 2nd Conference on Machine Translation, Proceedings (2017). https://doi.org/10.18653/v1/w17-4770
13. Post, M.: A call for clarity in reporting BLEU scores. In: Proceedings of the Third Conference on Machine Translation: Research Papers, pp. 186–191. Association for Computational Linguistics, Brussels (2018). https://www.aclweb.org/anthology/W18-6319
14. Rei, R., et al.: COMET-22: unbabel-IST 2022 submission for the metrics shared task. In: Proceedings of the Seventh Conference on Machine Translation (WMT), pp. 578–585. Association for Computational Linguistics, Abu Dhabi (2022). https://aclanthology.org/2022.wmt-1.52/
15. Rei, R., et al.: COMETKIWI: IST-unbabel 2022 submission for the quality estimation shared task. In: Proceedings of the Seventh Conference on Machine Translation (WMT), pp. 634–645. Association for Computational Linguistics, Abu Dhabi (2022). https://aclanthology.org/2022.wmt-1.60/
16. Seinen, T.M., et al.: Use of unstructured text in prognostic clinical prediction models: a systematic review (2022). https://doi.org/10.1093/jamia/ocac058
17. Shlyk, D., Groza, T., Mesiti, M., Montanelli, S., Cavalleri, E.: REAL: a retrieval-augmented entity linking approach for biomedical concept recognition. In: Demner-Fushman, D., Ananiadou, S., Miwa, M., Roberts, K., Tsujii, J. (eds.) Proceedings of the 23rd Workshop on Biomedical Natural Language Processing, pp. 380–389. Association for Computational Linguistics, Bangkok (2024). https://doi.org/10.18653/v1/2024.bionlp-1.29

18. Tsai, C.T., Upadhyay, S., Roth, D.: Multilingual Entity Linking. Springer, Cham (2024). https://doi.org/10.1007/978-3-031-74901-8
19. Upadhyay, S., Gupta, N., Roth, D.: Joint multilingual supervision for cross-lingual entity linking. In: Proceedings of the 2018 Conference on Empirical Methods in Natural Language Processing, EMNLP 2018 (2018). https://doi.org/10.18653/v1/d18-1270

Modeling Out-of-Vocabulary Words via Grammatical Fusion

Dror Mughaz

Lev Academic Center, Havaad Haleumi 21, 9116001 Jerusalem, Israel
myghaz@gmail.com

Abstract. Text embedding is a crucial element in computational linguistics, allowing computers to represent human language semantics numerically. A key approach is the Vector Space Model (VSM), which maps text into high-dimensional vectors and supports tasks like classification, clustering, and semantic search. Building on this, techniques like word embeddings (e.g., word2vec) and document embeddings have advanced Natural Language Processing (NLP), enabling sentiment analysis, machine translation, and summarization. However, challenges persist in morphologically rich languages (MRLs) such as Arabic, Hebrew, Russian, and Turkish. These languages feature complex structures with prefixes, suffixes, and roots, complicating text representation and increasing out-of-vocabulary (OOV) words. Addressing these challenges is essential for improving NLP's applicability across diverse linguistic contexts.

This study proposes a novel methodology to overcome existing limitations by utilizing probabilistic corpus decomposition. The corpus is processed in three forms: words are represented as wholes, decomposed into grammatical components, or broken down into letters. The Word2Vec algorithm is applied to this "triple corpus", generating vectors for words, their components, and letters, ensuring consistent semantic alignment across all forms. Experiments with English and Hebrew corpora demonstrate the effectiveness of this approach. The fused vectors of grammatical components achieve high-quality representations, often ranking first in cosine similarity, thereby reducing out-of-vocabulary (OOV) issues and expanding the lexicon. By improving the representation of morphological structures, this methodology enhances embeddings for both English and Semitic languages, advancing NLP capabilities in a variety of linguistic contexts.

Keywords: Word Embeddings · Natural Language Text · Grammatical Affixes · Multi-Vector Fusion · Morphologically Rich Languages · Out-Of-Vocabulary

1 Introduction

Text, the foundation of human communication, presents a challenge for computers: how to represent its meaning numerically. Computers struggle to grasp the intricacies of human language. The Vector Space Model (VSM) [42] and its offspring - word embeddings [30,38,44], sentence embeddings [7], and document embeddings [27] - are instrumental in bridging this gap. VSM emerges as a cornerstone technique in Natural Language Processing (NLP), Information Retrieval (IR), Text Classification (TC), Clustering, Recommendation Systems, Information Extraction (IE), Semantic Search, Topic Modeling, Plagiarism Detection, text and data mining [10,17,20,22,35,46] etc.'. VSM represents text as vectors in a high-dimensional space, where similar words and documents reside closer together. The VSM ability to represent text in a way that captures semantic relationships makes it a versatile tool across in various tasks and in particular, in NLP tasks. This allows computers to grasp semantic relationships based on word usage [42]. Word embeddings, like word2vec [30] and GloVe [38] take this a step further. By analyzing word context within a massive corpus, they capture not just a word's definition, but also it is meaning in different situations [30]. This empowers computers with tasks like identifying synonyms, understanding sentiment, and even generating human-quality text [38]. Sentence and document embeddings [7,27] extend this power further, encoding not just individual words, but also the structure and flow of ideas within sentences and entire documents - crucial for tasks like text summarization and machine translation [23].

The VSM serves as a foundational cornerstone in NLP, providing a basic technique for the representation and analysis of textual data [9]. Its origins can be traced back to the 1970s through the pioneering efforts of Salton and his colleagues [42], who recognized the necessity of a systematic method for representing documents and user queries to enhance IR efficiency. VSM addresses this challenge by constructing a high-dimensional space wherein each dimension corresponds to a distinct term in the vocabulary [6]. Subsequently, documents and queries undergo transformation into numerical vectors within this space, where the weight assigned to each dimension reflects the term's importance within the text. This weighting often employs methodologies like Term Frequency-Inverse Document Frequency (TF-IDF) [40]. One of the VSMs' strength lies in its capability to quantify semantic similarity between documents based on the geometric distance between their vectors in the space. Documents containing similar terms and concepts will have vectors positioned closer together, thus facilitating VSM's effectiveness in tasks such as document retrieval and TC [8].

The benefits of these embedding techniques are undeniable. They are the backbone of modern NLP applications, powering everything from chatbots and virtual assistants to sentiment analysis and automated document classification. As the volume and complexity of textual data explode, so too must the sophistication of these techniques.

Word embedding techniques commonly involve fitting vectors to entire words through the construction of neural network models. However, in these methodologies, certain words that fail to meet a predefined frequency threshold, known as out-of-vocabulary (OOV) words, do not have corresponding vector representations; this can be more challenge for morphological rich languages (MRL). (Example for MRL is Semitic languages such as Arabic, Aramaic, Amharic and

Hebrew, in which there is a root with a construction or weight; And Indian languages in which there is affixation of endings that skew the meaning of the word.) Consequently, these models do not encompass the entirety of human language, nor do they accommodate newly coined words or terms absent during the model's training phase. Consequently, when tasked with representing a sentence, issues may arise with models like word2vec and GloVe, as they lack familiarity with certain words within the sentence.

Several potential solutions have been proposed to address this challenge. The first method involves ignoring unknown words altogether, which is a straightforward approach. In this scenario, any word not present in the model's vocabulary is disregarded, and the sentence representation is based solely on the known words [25]. While this method can be effective when unknown words are rare, it may result in the loss of valuable information. Alternatively, another approach entails assigning a default vector, often a zero vector, to represent all unknown words [13]. This strategy ensures that every word contributes to the sentence representation, but it does not capture any semantic meaning associated with the unknown word.

The techniques of text embedding in data science hales with unveiling meaning from text. Although word embedding was the beginning, this field has developed a lot. These techniques offer a powerful solution, transforming individual words, sentences, paragraphs, and even entire documents into numerical vectors. These vectors capture the semantic relationships between these textual units, enabling machine learning models to effectively process and analyze vast amounts of textual data. Very important advancements in embedding techniques was the use of transformers that explored by [45] and [14]. The transformers have yielded significant progress in tasks like sentiment analysis, machine translation, and question answering. This section delves into the concept of text embedding, exploring various techniques used for embedding words, sentences, paragraphs, and documents.

While English, a dominant language in science and academia, belongs to the Indo-European family, Semitic languages like Aramaic and Hebrew present distinct characteristics: (1) Writing Direction: Unlike left-to-right English, Hebrew is written right-to-left [33]. (2) Morphological Complexity: Aramaic and Hebrew exhibit rich morphology, meaning words can have numerous prefixes that alter their meaning (e.g., "and when in...", "and when...", "and...", "when...", "in...") [18,19,21]. This complexity can lead to ambiguity in word interpretation. (3) Acronyms and Abbreviations: Hebrew texts frequently use acronyms and abbreviations [37]. (4) Historical Depth: As an ancient language spoken since the second millennium BC, Hebrew boasts a vast textual corpus, encompassing modern and ancient writings, with a significant portion of the ancient writings being religious texts from Rabbinic Judaism [31,32,36].

In this study, I propose a methodology for constructing a corpus that encompasses entire words, their grammatical components, and individual letters. Take the corpus and duplicate it three times (it is concatenated three times), then it undergoes probabilistic processing: each word has a certain likelihood of remaining intact, being decomposed into its grammatical components (prefix, stem, suffix), or being further broken down into letters. By triplicating the corpus in this manner, each word appears in three forms: as a whole; broken down into prefix, stem, suffix; and broken down into letters. This triple corpus structure ensures that each word, its grammatical components, and its letters shares the same contextual environment.

Applying the Word2Vec algorithm to this triple corpus yields vectors for whole words, prefix, stem, suffix, and letters. Because each word, regardless of its decomposition, shares the same contextual window, similar vectors are generated for specific words and their constituent parts. This direct assessment of vector quality is facilitated by the identical model applied across all forms of the corpus.

My contribution to both English and Hebrew languages lies in: Generating vectors for stem and grammatical affixes. Producing vectors representing entire words related to affixes. Enriching our lexicon by combining word parts to create new vocabulary i.e., minimizing OOV.

2 Related Work

In the past decade, I witnessed a surge in the development and application of vector embeddings, a technique that revolutionized Natural Language Processing (NLP). This technique facilitating the representation of textual data in a vector space which enables algorithms to capture semantic relationships between words and documents.

In 2013 Mikolov et al., [30] introduced word2vec a technique that use neural network which learns word representations by analyzing word co-occurrence within a massive corpus. Word2Vec goes through all the words serially, for each word the Distributional Semantics is a small window of words around the relevant word. In this way the neural network, map each word to a high-dimensional vector, where similar words reside closer together. This allows computers to grasp semantic relationships between words based on their usage patterns.

Following Le et al., [27], Pennington et al., [38] proposed the GloVe technique. GloVe also use neural network which learns word representations by leverages co-occurrence statistics from a large corpus to create word vectors. One of GloVe limitation is its computational complexity, as GloVe requires larger computational resources and longer training times [38] compared to Word2Vec, making it less efficient for large-scale datasets. Furthermore, GloVe embeddings are fixed and unable to adapt to varying contexts or domain-specific nuances, which can limit their effectiveness in specialized tasks or domains.

Word2Vec and GloVe are trained on a fixed vocabulary extracted from the training corpus. As a result, they may encounter out-of-vocabulary words during inference that were not present in the training data. Handling OOV words is challenging for these models, as they lack embeddings for such words, which can lead to difficulties in downstream NLP tasks.

In order to tackle with this challenge in 2017 Bojanowski et al., [11] developed the FastText algorithm. FastText represent vectors for subword, i.e., n-gram character. By representing words as bags of character n-grams, FastText can generate embeddings for rare and unseen words by leveraging their subword components. Additionally, Joulin et al., [24] explore, also, using character n-grams instead of words to represent text for classification with linear models. It shows that representing text as bags-of-character n-grams leads to faster training speeds compared to traditional bags-of-words.

In the following years, more advanced methods were developed; researchers have proposed several innovative solutions to overcome the challenges associated with vector embedding by using contextual embedding models like ELMO [39] and BERT [14]. These methods have gained prominence for their ability to generate word representations based on context, thus mitigating the issue of OOV words and capturing richer semantic information.

In their 2018 paper, Peters et al. propose Embeddings from Language Models (ELMo) [39], a groundbreaking method for generating word representations that consider context, i.e., deep contextualized word representations. Unlike static models like Word2Vec and GloVe, ELMo does not assign a single meaning to each word. Instead, it utilizes a bidirectional LSTM network to analyze a word's surroundings in a sentence, capturing both preceding and following words. This allows ELMo to grasp the syntactic and semantic nuances of word meaning, leading to substantial improvements in various NLP tasks. The authors highlight ELMo's effectiveness on benchmark datasets, demonstrating its superiority over static word embedding approaches in tasks like question answering, sentiment analysis, and named entity recognition.

Devlin et al., 2019 paper introduces BERT [14], a pioneering pre-training method for deep bidirectional transformers aimed at enhancing natural language understanding. BERT uses the more advanced Transformer architecture compared to ELMo's bidirectional LSTM. It addresses limitations of prior models through masked language modeling, where words are masked and the model predicts them based on surrounding context. Unlike prior left-to-right or right-to-left methods, BERT processes language bidirectionally and simultaneously. This allows it to leverage both preceding and following words for better contextual understanding. As a result, simply fine-tuning BERT achieves superior results to previous methods on 11 natural language processing tasks, such as question answering and language inference.

ELMo and BERT differ from previous word embedding approaches in that they produce contextualized representations of words dependent on the surrounding sentence, rather than static vectors for each word. Because word can take on different senses based on its context. Both ELMo and BERT address this by computing context-sensitive representations for each word token using Deep BiLMs/Transformers. As Peters et al., [39] explain regarding ELMo, "the representations produced by ELMo are context-dependent - not just functions of the words themselves, but also their surrounding context. This means that the same word may have different ELMo representations depending on how it is used in context". Devlin et al., [14] also note that BERT's masked language modeling pre-training objective enables context-dependent representations to be learned. Because of this context-awareness, if a word has the same meaning in two sentences, its specific context can lead to different vector representations.

The out-of-vocabulary (OOV) problem in vector representations like Word2Vec is more pronounced in MRL. MRL exhibit complex morphology, where words formed by combining morphemes, the general structure of a word in these languages is pre-stm-suf. The affixes are grammatical additions that express: time, gender, singular, plural, etc.' Word2vec, GloVe etc.' struggles with an affixes as it treats each word form as a separate entity. This can lead to a lack of meaningful representations for words derived from known roots [11,12,29].

The methods Word2Vec, GloVe, and FastText are language-independent, allowing for the creation of vectors either independently or by utilizing pre-existing vectors available online. Vectors tailored for the English language can be found online for each of the three methods. Vectors for Modern Hebrew can be sourced from the internet; however, vectors for ancient Hebrew necessitate independent construction.

Studies focusing specifically on Hebrew are much less common. Mughaz et al., [34] investigated the writer's opinion about products by classifying short reviews written in Hebrew. Their approach involved parsing reviews into sentences, utilizing unigrams and bigrams and employing SVMs with feature selection for classification. This method achieved a success rate of over 92%. Liebeskind et al., [28] explored sentiment classification for Hebrew Facebook posts from politicians. They compared nine machine-learning techniques and two sentiment classification tasks (general attitude and content-specific attitude). Their results showed that the Logistic Regression method exceeded the other eight ML models in terms of F-measures and accuracy.

3 Data

I conducted tests on two corpora: one in English and the other in Hebrew. Before generating vectors using Word2Vec, I applied the following two preprocessing steps to each corpus:

Corpus Triplication - Each corpus was replicated three times by concatenating it with itself, resulting in what I call "the triple corpus", where the same text appears three times.

Word Representation Variability - A word could be represented in three different ways: (1) as a whole word; (2) segmented into its grammatical components (prefix, stem, and suffix); or (3) broken down into individual letters. Each word in the corpus was processed with a certain probability of remaining whole, another probability of being decomposed into its grammatical components, and another probability of being further divided into letters. The probability calculations will be presented later.

This triplication approach ensures that each word appears in three forms- whole, segmented by grammatical structure, and split into letters-while maintaining a shared contextual environment.

3.1 Comparison of English Text Size and Hebrew Text Size

To compare the sizes of "identical" text in Hebrew and English, I used Yandex Translate [5]. I translated Hebrew text into English and English text into Hebrew. The size of the texts was approximately 5MB, the results appear in Table 1.

Table 1. Comparison between Hebrew and English

	Eng	Heb	Eng/Heb Comparison
AVG word length	~5.12 chars	~4.88 chars	~1.05
Total #words	~10.1M words	~7.5M words	~1.35
AVG #occurrences per word	~19.2	~10.5	~1.8

Conclusions:

- In terms of words, English text is 1.4 times larger (1.35 * 1.05) than text in Hebrew → affects the text size.
- The number of different words in Hebrew is 1.8 times greater in English → affects the size of the min_count.

The Experimented Texts. As previously mentioned, the original text was decomposed to create a new file, which I used to build the word2vec vectors. This new file contains: (1) whole words, (2) words divided into linguistic elements, and (3) words broken into letters. To ensure no information was lost, I tripled the source file.

English Corpus: The English corpus consists of Wikipedia texts [2], with a total size of 1.5GB. To construct the text optimally, I used a spaCy stemmer [1] to obtain various word decompositions.

- Each word breaks down to an average of 1.17 sub-words.
- 6.3% of words have a linguistic decomposition.
- The average length of each word is 4.3 letters (a word is also a semicolon, etc.).

According to the above statistics, the breakdown of words in the text is as follows: 47.9% of the words remained intact, 40.9% were decomposed into linguistic elements, and 11.2% were broken down into letters.

Hebrew Corpus: The Hebrew corpus comprises Wikipedia texts, along with articles from Walla and Ma'ariv [3], totaling 1.1GB. To construct the text optimally, I used the Sade et al. stemmer [41] to obtain various word decompositions.

- Each word breaks down to an average of 1.46 sub-words.

- 40% of words have a linguistic decomposition.
- The average length of each word is 4 letters (a word is also a semicolon, etc.).

According to the above statistics, the breakdown of words in the text is as follows: 51.7% of the words remained intact, 35.3% were decomposed into linguistic elements, and 13% were broken down into letters. Thus, in both corpora, one-third of the tokens in the tripled corpus are complete words, one-third are decomposed into linguistic elements, and one-third are letters. This structure allows us to build a single model that includes words, linguistic elements, and letters. It might seem that these percentages (47.9%, 40.9%, 11.2% for English; 51.7%, 35.3%, 13% for Hebrew) do not align with: "one-third are decomposed into linguistic elements, and one-third are letters". However, breaking a word into its linguistic elements or individual letters significantly increases the number of tokens.

3.2 Embedding the English Tokens

I use word2vec [30] (on the new file) to convert the words into vectors for that I used the genism [4] tool with the following hyper parameters:

- The size of the vector is 300
- The minimum appearance of the word (to build for it vector) is 35.
- The window size around the learned word is 14 instead of 10 (Adjustments as a result of the changes I made to the source file)
- The function to build the vectors is skip-gram

I ended with 274,519 vectors/tokens, of these, 265,077 are complete words (not parts of words or letters).

3.3 Embedding the Hebrew Tokens

I use word2vec [30] (on the new file) to convert the words into vectors for that I used the genism [4] tool with the following hyper parameters:

- The size of the vector is 300
- The minimum appearance of the word (to build for it vector) is 18.
- The window size around the learned word is 16 instead of 10 (Adjustments as a result of the changes I made to the source file)
- The function to build the vectors is skip-gram

I ended with 279,279 vectors/tokens, of these, 270,789 are complete words (not parts of words or letters).

4 Experiment and Results

After constructing the model as described, the model includes vectors representing words, both in terms of their linguistic components and the individual letters that form them.

My hypothesis is that a word can be constructed from its parts to produce a well-representative vector. An example of constructing a word by assembling its components:

Unlike = Un+like = U+n+like = Un+l+i+k+e = U+n+l+i+k+e To test my hypothesis, I followed these steps:

1. I selected a word with known linguistic components and obtained its vector representation.
2. I extracted the vectors representing the linguistic components of that word.
3. I created a composite vector by summing the vectors of these linguistic components.
4. I compared the original word vector from step 1 with the composite vector from step 3, analyzing two outcomes: the cosine similarity between the vectors and the position of the composite vector relative to all other word and component vectors.

(I also repeated these steps using the individual letters and linguistic components of the words).

4.1 Results

The following tables present the results of the two experiments. Table 2 displays the results for the English language, while Table 3 shows the results for the Hebrew language. In Table 3, I have included English translations of the Hebrew words.

4.2 Results Discussion

The results in Tables 2 and 3 suggest that the vector closest to the one made up of the prefix, stem, and suffix is usually the most accurate representation of the complete word, apart from the original vector itself. For instance, the vector representing the combination 'un + like' is the nearest to the vector for the word 'unlike'. In the case of a Hebrew word, the vector for נקוד+ות is the closest to the vector for נקודות. I assume this method will be good for other languages as well.

Currently, when there is no representative vector for a word, users of Word2vec or GloVe resort to using either a random vector or a zero vector. The average distance between these placeholders and the actual vector needed corresponds to the number of vectors divide by 2. Constructing a vector using the morpheme vectors yields significantly better results, as demonstrated in Tables 2 and 3, outperforming the random or zero vectors, which is around 132,538 for English and 135,394 for Hebrew. When constructing a vector from individual

Table 2. Comparison of English word composition, cosine similarity, and the position of the composed vector relative to the complete word vector.

word target	token assembling	Cosine similarity	Place
unlike	Un+like	0.53	1
	U+n+like	0.34	22
	Un+l+i+k+e	0.15	1622
	U+n+l+i+k+e	0.12	5339
being	Be+ing	0.55	1
	B+e+ing	0.24	6087
	Be+i+n+g	0.15	29308
	B+e+i+n+g	0.07	260008
returned	Return+ed	0.56	1
	Re+turn+ed	0.55	1
	Re+turned	0.37	32
	Re+turn+e+d	0.32	30
	R+e+turn+ed	0.29	197
	R+e+turn+e+d	0.19	18824
	R+e+t+u+r+n+e+d	0.09	271503
unusual	Un+usual	0.7	1
	U+n+usual	0.46	77
	Un+u+s+u+a+l	0.19	21241
	U+n+u+s+u+a+l	0.15	63147

Table 3. Comparison of Hebrew word composition, cosine similarity, and the position of the composed vector relative to the complete word vector.

translation	word target	token assembling	Cosine similarity	Place
that no	שלא	ש+לא	0.79	1
		ש+ל+א	0.44	9
points	נקודות	נקוד+ות	0.72	1
		נקוד+ו+ת	0.41	13
		נ+ק+ו+ד+ות	0.13	9334
the last ones	האחרונים	האחרון+ים	0.55	1
		ה+אחרון+ים	0.39	13
		ה+אחרונים	0.19	841
		ה+אחרון+י+ם	0.07	43835
the years	השנים	ה+שנה+ים	0.46	1
		ה+שנים	0.5	3
		השנה+ים	0.28	92
		ה+ש+נ+ים	0.11	48631
		ה+ש+נ+י+ם	0.08	103105

letter vectors, the results are less favorable. However, even these letter-based vectors generally perform better than random or zero vectors.

It is important to emphasize, grammatical additions like prefixes and suffixes are integral to standard sentences, making their occurrence very common. Because these grammatical elements appear frequently in the original corpus, they will also be prevalent in the "triple" corpus as separate units. Their frequent presence in relevant sections of the text ensures they have a good vector representation.

5 Future Works

Today, more advanced NLP tools are available, which I believe can achieve better results than those presented in this paper, especially for morphologically rich languages like Hebrew. Morphological analysis is not an independent NLP task but a crucial part of larger tasks. Therefore, to improve results, it is essential to use advanced NLP tools that include a morphological analysis phase. The overall task results can be used indirectly to evaluate the improvement. Lemmatization, which relies heavily on identifying the constituent morphemes within a word, is one such task suitable for this assessment.

I used a tool developed in 2018 [15]. At that time, the success rate of the lemmatization task for Hebrew was 75.67% [16]; today, it has increased to 95.58% [26]. For English, the success rate of the lemmatization task was 97.03%–97.53% [43], and today it ranges from 97.68% to 99.06% [1]. I observe an improvement in the lemmatization task for both English and Hebrew. In English, the improvement ranges from 0.7% to 2.1%, while in Hebrew, it is significantly higher at 26.3%. These results support my assessment that better outcomes are achievable.

I plan to use newer tools to achieve better results, particularly for the Hebrew language. I intend to apply the same process to other linguistically rich languages as well. Additionally, I plan to incorporate the aforementioned vectors in the embedding layer of neural networks.

References

1. spacy stemmer. https://stackabuse.com/python-for-nlp-tokenization-stemming-and-lemmatization-with-spacy-library/
2. English Wikipedia (2019). https://dumps.wikimedia.org/enwiki/
3. Hebrew Wikipedia (2019). https://dumps.wikimedia.org/hewiki/
4. Word2vec embeddings (2019). https://radimrehurek.com/gensim/models/word2vec.html
5. Yandex translate (2022). https://translate.yandex.com/
6. Abualigah, L., et al.: Efficient text document clustering approach using multi-search arithmetic optimization algorithm. Knowl. Based Syst. **248**, 108833 (2022)
7. Arora, S., Liang, Y., Ma, T.: A simple but tough-to-beat baseline for sentence embeddings. In: International Conference on Learning Representations (2017)

8. Baeza-Yates, R.: Modern Information Retrieval, vol. 2, pp. 127–136. Addison Wesley (1999)
9. Bansal, S.: Vector representation of documents using word clusters. Ph.D. thesis, Concordia University (2021)
10. Berge, G.T., Granmo, O.C., Tveit, T.O., Ruthjersen, A.L., Sharma, J.: Combining unsupervised, supervised and rule-based learning: the case of detecting patient allergies in electronic health records. BMC Med. Inform. Decis. Mak. **23**(1), 188 (2023)
11. Bojanowski, P., Grave, E., Joulin, A., Mikolov, T.: Enriching word vectors with subword information. Trans. Assoc. Comput. Linguist. **5**, 135–146 (2017)
12. Botev, G., McCarthy, A.D., Wu, W., Yarowsky, D.: Deciphering and characterizing out-of-vocabulary words for morphologically rich languages. In: Proceedings of the 29th International Conference on Computational Linguistics, pp. 5309–5326 (2022)
13. Che, X., Luo, S., Yang, H., Meinel, C.: Sentence boundary detection based on parallel lexical and acoustic models. In: Interspeech, pp. 2528–2532 (2016)
14. Devlin, J., Chang, M.W., Lee, K., Toutanova, K.: BERT: pre-training of deep bidirectional transformers for language understanding. In: Proceedings of the 2019 Conference of the North American Chapter of the Association for Computational Linguistics: Human Language Technologies, Volume 1 (Long and Short Papers), pp. 4171–4186 (2019)
15. Dumitrescu, Ş.D., Boroş, T., Tufiş, D.: RACAI's natural language processing pipeline for universal dependencies. In: Proceedings of the CoNLL 2017 Shared Task: Multilingual Parsing from Raw Text to Universal Dependencies, pp. 174–181 (2017)
16. Eyal, M., Noga, H., Aharoni, R., Szpektor, I., Tsarfaty, R.: Multilingual sequence-to-sequence models for Hebrew NLP. arXiv preprint arXiv:2212.09682 (2022)
17. Gupta, A., Goyal, R.: A generative AI-driven method-level semantic clone detection based on the structural and semantical comparison of methods. IEEE Access (2024)
18. HaCohen-Kerner, Y., Beck, H., Yehudai, E., Mughaz, D.: Identifying historical period and ethnic origin of documents using stylistic feature sets. In: International Conference on Discovery Science, pp. 102–113. Springer (2006)
19. HaCohen-Kerner, Y., Beck, H., Yehudai, E., Mughaz, D.: Stylistic feature sets as classifiers of documents according to their historical period and ethnic origin. Appl. Artif. Intell. **24**(9), 847–862 (2010)
20. HaCohen-Kerner, Y., Mughaz, D.: Estimating the birth and death years of authors of undated documents using undated citations. In: International Conference on Natural Language Processing, pp. 138–149. Springer (2010)
21. HaCohen-Kerner, Y., Mughaz, D., Beck, H., Yehudai, E.: Words as classifiers of documents according to their historical period and the ethnic origin of their authors. Cybern. Syst. Int. J. **39**(3), 213–228 (2008)
22. HaCohen-Kerner, Y., Schweitzer, N., Mughaz, D.: Automatically identifying citations in Hebrew-Aramaic documents. Cybern. Syst. Int. J. **42**(3), 180–197 (2011)
23. Hailu, T.T., Yu, J., Fantaye, T.G.: A framework for word embedding based automatic text summarization and evaluation. Information **11**(2), 78 (2020)
24. Joulin, A., Grave, E., Bojanowski, P., Mikolov, T.: Bag of tricks for efficient text classification. arXiv preprint arXiv:1607.01759 (2016)
25. Kandi, S.M.: Language modelling for handling out-of-vocabulary words in natural language processing. Diss. Doctoral dissertation (2018)
26. Kondratyuk, D., Gavenčiak, T., Straka, M., Hajič, J.: LemmaTag: jointly tagging and lemmatizing for morphologically-rich languages with BRNNs. arXiv preprint arXiv:1808.03703 (2018)

27. Le, Q., Mikolov, T.: Distributed representations of sentences and documents. In: International Conference on Machine Learning, pp. 1188–1196. PMLR (2014)
28. Liebeskind, C., Nahon, K., HaCohen-Kerner, Y., Manor, Y.: Comparing sentiment analysis models to classify attitudes of political comments on Facebook (November 2016). Polibits **55**, 17–23 (2017)
29. Luong, M.T., Socher, R., Manning, C.D.: Better word representations with recursive neural networks for morphology. In: Proceedings of the Seventeenth Conference on Computational Natural Language Learning, pp. 104–113 (2013)
30. Mikolov, T.: Efficient estimation of word representations in vector space. arXiv preprint arXiv:1301.3781 (2013)
31. Moghaz, D., Hacohen-Kerner, Y., Gabbay, D.: Text mining for evaluating authors' birth and death years. ACM Trans. Knowl. Disc. Data (TKDD) **13**(1), 1–24 (2019)
32. Mughaz, D.: Classification of Hebrew texts according to style. Unpublished master's thesis [in Hebrew], Bar-Ilan University, Ramat-Gan, Israel (2003)
33. Mughaz, D., Cohen, M., Mejahez, S., Ades, T., Bouhnik, D.: From an artificial neural network to teaching. Interdisc. J. e-Skills Lifelong Learn. **16**, 001–017 (2020)
34. Mughaz, D., Fuchs, T., Bouhnik, D.: Automatic opinion extraction from short Hebrew texts using machine learning techniques. Comput. Sist. **22**(4), 1347–1357 (2018)
35. Mughaz, D., HaCohen-Kerner, Y., Gabbay, D.: Mining and using key-words and key-phrases to identify the era of an anonymous text. In: Transactions on Computational Collective Intelligence XXVI, pp. 119–143. Springer (2017)
36. Mughaz, D., HaCohen-Kerner, Y., Gabbay, D.: Extracting and tagging unstructured citation of a Hebrew religious document. In: InSITE 2019: Informing Science+ IT Education Conferences: Jerusalem, pp. 461–473 (2019)
37. Mughaz, D., HaCohen-Kerner, Y., Gabbay, D.: Extraction of time-related expressions using text mining with application to Hebrew. PLoS ONE **19**(2), e0293196 (2024)
38. Pennington, J., Socher, R., Manning, C.D.: GloVe: global vectors for word representation. In: Proceedings of the 2014 Conference on Empirical Methods in Natural Language Processing (EMNLP), pp. 1532–1543 (2014)
39. Peters, M.E., et al.: Deep contextualized word representations. In: Walker, M., Ji, H., Stent, A. (eds.) Proceedings of the 2018 Conference of the North American Chapter of the Association for Computational Linguistics: Human Language Technologies, Volume 1 (Long Papers), pp. 2227–2237. Association for Computational Linguistics, New Orleans (2018). https://doi.org/10.18653/v1/N18-1202, https://aclanthology.org/N18-1202
40. Rahardi, M., Aminuddin, A., Abdulloh, F.F., Nugroho, R.A.: Sentiment analysis of COVID-19 vaccination using support vector machine in Indonesia. Int. J. Adv. Comput. Sci. Appl. **13**(6) (2022)
41. Sade, S., Seker, A., Tsarfaty, R.: The Hebrew universal dependency treebank: past present and future. In: Proceedings of the Second Workshop on Universal Dependencies (UDW 2018), pp. 133–143 (2018)
42. Salton, G., Wong, A., Yang, C.S.: A vector space model for automatic indexing. Commun. ACM **18**(11), 613–620 (1975)
43. Toporkov, O., Agerri, R.: On the role of morphological information for contextual lemmatization. Comput. Linguist. **50**(1), 157–191 (2024)
44. Treistman, A., Mughaz, D., Stulman, A., Dvir, A.: Word embedding dimensionality reduction using dynamic variance thresholding (DYVAT). Expert Syst. Appl. **208**, 118157 (2022)

45. Vaswani, A., et al.: Attention is all you need. In: Advances in Neural Information Processing Systems, vol. 30, pp. 5998–6008 (2017)
46. Zhang, Z., et al.: Scholarly recommendation systems: a literature survey. Knowl. Inf. Syst. **65**(11), 4433–4478 (2023)

Early Length of Stay Prediction at Admission in Short-Stay Hospitals

Mohamed Gharbi[1,2,3(✉)], Christine Verdier[2], Maria Di Mascolo[3], and Jean-Marc Babouchkine[1]

[1] Calystene, Eybens, France
mohamed.gharbi1@univ-grenoble-alpes.fr
[2] Univ. Grenoble Alpes, CNRS, Grenoble INP, LIG, 38000 Grenoble, France
christine.verdier@univ-grenoble-alpes.fr
[3] Univ. Grenoble Alpes, CNRS, Grenoble INP, G-SCOP, 38000 Grenoble, France
maria.di-mascolo@grenoble-inp.fr

Abstract. Predicting hospital length of stay (LOS) is crucial for optimizing resource allocation and improving patient care. Accurate estimation of a patient's discharge date enhances operational efficiency and reduces stress on hospital staff. While many studies have addressed this challenge using classification and regression models that consider data from admission to discharge, this research focuses on predicting LOS using only the features available at admission, specifically demographic details and primary diagnosis data, in a case study of patients in surgical wards dealing with musculoskeletal and connective tissue conditions. Advanced machine learning techniques, including boosting algorithms (LightGBM, XGBoost, CatBoost), Random Forest, and Feedforward Neural Networks (FNN), were employed to achieve this. LightGBM, demonstrated the best performance in this context, achieving a precision R^2 score of 0.77 and a mean squared error (MSE) of 1.07. These measures indicate the model's ability to predict LOS with a high degree of accuracy while minimizing prediction errors.

These results highlight the potential of machine learning, particularly LightGBM, to optimize hospital operations and enhance patient outcomes. Ultimately, the findings underscore the value of leveraging admission-based data for early and accurate LOS predictions, providing a foundation for more efficient resource planning in healthcare settings.

Keywords: Machine learning · Electronic Medical Records · Length Of Stay Prediction · Regression

1 Introduction

The length of stay (LOS) is widely recognized as a key metric for assessing hospital performance and operational efficiency, as emphasized by the World Health Organization (WHO) [1]. Defined as the duration of a patient's hospitalization for a single episode of care, LOS provides critical insights into the effectiveness of healthcare delivery. The LOS can vary significantly across different types of hospitalization. For instance, short-stay hospitalizations, which include medical, surgical, and obstetric care, typically have a mean LOS of 6 days. These stays are characterized by unpredictability, as factors such as the patient's state, diagnosis, comorbidity, and individual characteristics can all influence the length of hospitalization. Efficient management of LOS is crucial across managerial, medical, and financial domains, as it directly impacts resource allocation and patient outcomes. Conversely, delays in discharge or prolonged stays can disrupt workflows, strain resources, and negatively affect the experiences of both patients and staff [2]. Therefore, there is a need for accurate prediction of LOS in short-stay hospitalization to address these issues.

Predicting the length of stay (LOS) can significantly benefit hospitals by enabling more informed medical decision-making and efficient planning. Accurate LOS predictions allow hospitals to provide families with an estimated discharge date, which helps manage expectations. Furthermore, such predictions facilitate the advanced mobilization of resources, including ensuring an adequate number of nursing staff to address workload distribution challenges. Research has demonstrated the value of early LOS prediction in improving hospital operations. For instance, studies such as [3] have shown that predicting LOS can help reduce costs, enhance healthcare delivery, and enable the timely mobilization of critical resources, such as ICU beds.

Many studies have been conducted to predict LOS in short-stay hospitals, leveraging various types of information, such as demographic, clinical, and administrative data. However, most of these studies rely on data available up to the point of discharge to predict LOS. In contrast, our study aims to predict LOS using only the data available at admission.

In this study, we utilized Electronic Medical Records (EMRs), which are digital systems used by hospitals to track patients' medical history, treatments, and demographic information. Data extracted from EMRs were employed to predict the length of stay (LOS) in short-stay hospitalizations. To achieve this, we implemented several machine learning algorithms, including ensemble models such as LightGBM, Random Forest (RF), CatBoost, and XGBoost, which are well-suited for capturing complex patterns and relationships between different types of variables while effectively handling noise [4]. Additionally, we tested Feedforward Neural Networks (FNNs) for their robustness in modeling intricate patterns and extracting insights from various variables.

This paper is organized as follows: Sect. 2 reviews the related work, highlighting existing approaches and methodologies for predicting the length of stay (LOS) in hospitals. Section 3 introduces the case study that serves as the basis for our research, describing the hospital setting, and the characteristics of the dataset used. Section 4 outlines our proposed approach for LOS prediction, detailing the methods and techniques employed. This section is further divided into two subsections: Sect. 4.1, which explains the data processing steps, including cleaning, transformation, and feature selection, and Sect. 4.2, which describes the machine learning models and algorithms implemented to predict

LOS. Section 4.3 presents the evaluation metrics and results, comparing the performance of different models and discussing their implications. Finally, the Conclusion summarizes the main findings, discusses the limitations of our work, and proposes directions for future research.

2 Related Work

Many studies have highlighted the role of artificial intelligence (AI) in the healthcare domain [5], particularly in resource management, to address unpredictable situations such as hospital overcrowding, surgery cancellations or delays, and workload management. These challenges can exacerbate difficulties in hospital operations. One solution that has been explored is the prediction of hospital length of stay (LOS) [6].

LOS prediction has been applied across diverse clinical contexts, including lung cancer, neonatal intensive care units (NICUs), ankle arthroplasty, bed census forecasting, and patients undergoing coronary artery bypass graft (CABG) surgery. A variety of machine learning algorithms have been employed, ranging from classification methods and ensemble models to neural networks (NN), with a primary focus on classification tasks. The most commonly used models were k-nearest neighbors (KNN), Random forest (RF), Decision Tree, Support Vector Machine (SVM), Linear Regression (LR), XGBoost, and neural networks (NNs). For instance, a study reported that the random forest classification algorithm achieved the best performance with an accuracy of 97% for binary length-of-stay (LOS) prediction in NICUs. In this study, the length of stay was categorized as a short stay (less than 3 days) or a long stay (more than 3 days) [7].

A recent review conducted in 2024 [6], compared various articles presenting machine learning methods for predicting hospital length of stay (LOS) using both classification and regression approaches. The review found that neural networks achieved superior performance in LOS prediction, while traditional machine learning methods also demonstrated good results, though their effectiveness was highly dependent on dataset characteristics. Classification was the most commonly used approach, with LOS often categorized into intervals.

Some studies have also focused on predicting LOS at admission, leveraging historical data that includes patient characteristics and treatment details up to discharge. However, relatively few studies have explored LOS prediction using only admission data [8]. One of the articles that showcased this is [9], where the authors predicted the LOS of readmitted patients at admission using demographic information and data of the first admission in the classification approach.

In another study [10], the authors applied machine learning models for predicting continuous LOS, using several features at admission for programmed patients admission such as age, primary and secondary diagnosis, comorbidity, and laboratory data. Table 1, summarizes previous works, highlighting the hospital's ward/field, algorithms, used approaches, the number of features considered (N.F), and the specific scenarios for prediction.

Based on the analysis in Table 1, most studies employed machine learning algorithms for classification purposes. Specifically, studies such as [7,9,11]

Table 1. Overview of articles in predicting Length of stay in hospitals. ***Abbreviations:*** *Ar - Article, N.F - Number of Features, Perf - Performance, ICU - Intensive Care Unit, LGB - Light Gradient Boosting, UMLS - Unified Medical Language System, Cls. - Classification, Reg. - regression.* ***Performance metrics:*** *Acc - Accuracy, R^2 and MAE - Mean Absolute Error*

Ar	Ward/ Field	Scenario	Algorithm	Approach	N.F	Perf
[7]	Neonatal	Discharge	RF	Cls. (Acc)	11	97%
[9]	ICU	Admission	RF	Cls. (Acc)	11	71%
[10]	ICU	Admission	LR and RF	Reg. (R^2)	13	36%
[11]	ICU	Discharge	RF	Cls. (Acc)	70	96%
[12]	Clinical ward	Discharge	RF and UMLS	Cls. (Acc)	10	75%
[13]	hospital	Admission	LGB and RF	Cls. (Acc)	9	81%
[14]	ICU	Admission	Ridge and Lasso	Reg. (MAE)	12	0.96
[15]	cardio-thoracic	Preoperative	Bayesian	Reg. (MAE)	12	1.16
[16]	ICU	Admission	Deep attention	Cls. (Acc)	8	81%

focused on binary classification for predicting the length of stay (LOS) at admission and discharge. In contrast, other works, including [12, 13, 16], addressed LOS prediction as a multi-classification problem by categorizing LOS into three distinct classes.

A smaller subset of studies [10, 14, 15], explored LOS as a regression problem. These studies employed a variety of approaches, from traditional machine learning to advanced deep learning techniques, with most focusing on data available up to the point of patient discharge. An exception was observed in [15], where the dataset included patient characteristics, the first laboratory results, the number of diagnoses, and comprehensive demographic details.

In conclusion, most studies on Length of Stay (LOS) prediction in hospitals have relied on historical data that captures patients' states, characteristics, and treatments, aiming to predict LOS at both admission and discharge. Exceptions include studies such as [13, 15], where models incorporated admission data alongside prescriptions specific to the patient's diagnosis. These studies highlight the potential of leveraging targeted data for more precise LOS predictions. A wide range of data sources were used in these studies, with Electronic Medical Records (EMRs) serving as the primary source. Open-source datasets, like MIMIC [17], were extensively utilized in studies such as [9, 15], underscoring their significance in LOS prediction research.

While most existing research has focused on the entire patient's hospitalization process, from admission to discharge, some studies, such as [15], primarily rely on admission data, supplemented by the patient's associated diagnosis and initial laboratory results. However, many of these approaches incorporate features that are available only after admission, such as follow-up laboratory tests or a comprehensive medical history, limiting their applicability in early decision-making.

In contrast, our study aims to predict the length of stay (LOS) using only the data available at the time of patient admission, including the date of birth, gender, principal diagnosis, and the motivation for admission, particularly for planned admissions. By relying solely on admission-specific data, our model offers practical insights that enable hospitals to optimize resource management, including forecasting nurse workload, before additional clinical information becomes available.

Our work fills a critical gap in current research by demonstrating that LOS can be accurately predicted without relying on post-admission data, which primarily provides retrospective insights. Instead, our approach focuses on prospective prediction, enabling real-time decision-making. This is particularly valuable for short-stay patients, where early planning is essential for efficient hospital operations.

3 Case Study

Electronic Medical Records (EMR) in hospitals contain a vast amount of valuable information that can be leveraged for hospital management and workload prediction. While there are already numerous research efforts in the field of AI in healthcare utilizing existing data, EMRs remain a rich source of information to address the needs of healthcare providers further and improve analysis.

For this study, we utilized structured data from the EMRs maintained by Futura Smart Design [18], a system developed by Calystene. The dataset encompasses patient demographics, clinical diagnosis, medical procedures, and recorded treatments throughout hospitalization.

Upon analyzing the EMR, we identified a column labeled 'provisional date of discharge,' which represents an estimated date provided by staff to approximate when a patient might be discharged. However, this column is frequently incomplete or missing. To address this gap, we aim to predict the length of stay (LOS) in hospitals at the time of patient admission using the data available at that point. Accurately predicting LOS is not only crucial for optimizing hospital occupancy management but also serves as a key component for estimating nursing workload. Since nursing workload is influenced by both the LOS and the intensity of treatments administered, LOS prediction provides the first step for future workload forecasting models.

Our research focuses specifically on predicting LOS per ward. For this study, we targeted the Surgical department by focusing on one family of diseases: Diseases of the musculoskeletal system and connective tissue as a case study, where patients stay and treatment plans often vary significantly. By understanding the LOS trends, we can begin to establish a framework for connecting LOS with nursing workload, an aspect we plan to explore in future work.

The following section outlines our proposed approach for predicting Length of Stay.

4 Proposed Approach

Prior to this work, we proposed a structured approach that generalized the process of nursing workload prediction [19], based on the Cross-Industry Standard Process for Data Mining (CRISP-DM) methodology [20]. This methodology consists of six steps in a data mining project: business understanding, data understanding, data processing, modeling, evaluation, and deployment.

In this study, we focus on two key components: data processing and modeling for predicting the length of stay at the admission stage, which forms part of the primary goal of forecasting nursing workload in short-stay hospitalizations.

4.1 Data Processing

This part of the process is composed of several steps: data collection, feature selection, and data cleaning.

Data Collection. As previously mentioned, we extracted data from the EMR system of the company, which includes five years of short-stay hospitalization records from 2008 to 2013. The dataset comprises a total of 8152 rows. These records contain information spanning from patient admission to discharge. However, since our objective focuses on using data available at the time of admission, we retained only features such as date of birth (DOB), gender, primary diagnosis (PD), associated diagnoses (AD), and their corresponding labels (PDL and ADL), as well as weight, reason for admission (RA), provisional date of entry (PDE), real date of entry (RDE), provisional date of discharge (PDD), and actual date of discharge (ADD).

In hospitals, diagnoses are coded using the ICD-10 (International Statistical Classification of Diseases and Related Health Problems, 10th Revision) system, which classifies diseases and health conditions into 22 main categories and their subgroups [21].

As discussed in Sect. 3, the provisional date of discharge is frequently unavailable. To support our objective of predicting the length of stay (LOS), we included the real date of discharge in the dataset for training and evaluation purposes.

Dataset Features Selection. At the patient admission stage, the available data is limited, and its availability varies depending on the type of hospitalization (short, medium, and long) and the ward. In our case, we used data from the surgical ward in short-term hospitalization. After careful discussion with the company's experts, we made an initial selection of features, as described in Sect. 4.1.

Subsequently, we made a second selection based on the data available in our dataset. We found that the variables weight, reason for admission, and secondary diagnosis were not reported for more than 85% of the cases. Additionally, the primary diagnosis labeling showed redundancy in the data. As a result, we retained only the variables age, gender, primary diagnosis codes, and length of stay.

Table 2, provides insight into the percentage of the missing values (MV) in each feature and its description.

Table 2. Features Description and Percentage of Missing Values (MV)

Feature	Description	MV (%)
RA	The primary reason for the patient's hospitalization	100
PED	The planned admission date for the patient, relevant for scheduled admissions	99.98
AD	Additional medical conditions diagnosed alongside the primary diagnosis	98.64
ADL	Label for the associated diagnoses	98.64
Weight	The weight of the patient (in kilograms)	86.3
DOB	The patient's date of birth	0
RDE	The actual date of the patient's hospital admission	0
Age	The patient's age (calculated from the DOB)	0
Gender[a]	The biological sex of the patient	0
PD	The primary diagnosis assigned to the patient upon admission	0
PDL	Label of the primary diagnosis	0
RDD	The actual date of the patient's hospital discharge	0
LOS	The total length of the patient's stay in the hospital (calculated from admission to discharge)	0

[a] In our dataset, gender refers to the biological sex of the patient, which was either male or female.

Data Cleaning. Working with raw data can be challenging and may impact model performance. To address this, we applied several preprocessing rules specific to the surgery ward. These included retaining only records with a length of hospitalization greater than 1 day and less than 15 days, as the mean length of stay (LOS) in short-stay hospitalizations is approximately six days [22].

Errors Detection. Several quality control measures were implemented to ensure the integrity of the data: the validation of admission dates; any admission date that was before the date of birth or after the date of discharge was removed, as such records were deemed invalid; removal of invalid diagnoses. Finally, incomplete patient records were excluded, specifically those missing essential information such as date of birth, date of admission, date of discharge, or gender. These fields were considered mandatory, and their absence was attributed to issues in the Electronic Medical Records (EMR) system or incompletion.

Diagnosis Feature Processing. To analyze the distribution, total number of diagnoses, and hierarchical levels used in the diagnosis, we processed the ICD-10 codes as follows:

- Aggregation of Low-Frequency Categories: Diagnoses with a frequency below 0.3% and above 0.1% were aggregated into their next hierarchical level to ensure adequate representation in the dataset. For instance, if "A047" had a frequency below 0.2%, it was assigned to the broader category "A04".
- Exclusion of Rare Diagnoses: Diagnoses that appeared less than 0.1 in the dataset were excluded from the analysis to focus on more prevalent cases. In addition, diagnoses that didn't follow the ICD-10 codification were deleted.
- Standardization of Code Length: Diagnoses were standardized to include up to four characters (e.g., "A047" instead of "A0471"), maintaining a meaningful detail of the diagnosis.

The results of our data preprocessing revealed that we have a dataset comprising 7880 patients with verified and reliable information, focusing on 42 unique diagnoses. The age of patients ranges from 2 to 96 years. Figure 1 illustrates the interaction of three key features-diagnosis, age, and gender-with the length of stay.

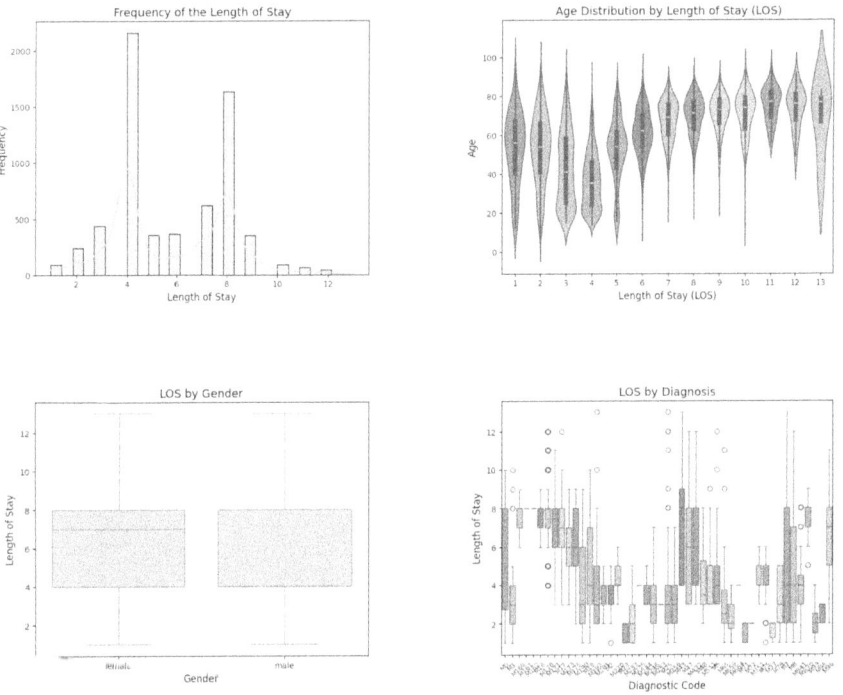

Fig. 1. Interaction of features with length of stay.

Our analysis indicates that gender does not significantly influence the length of stay, as no observable variability was noted based on this feature. In contrast, both diagnosis and age demonstrate a notable influence on the length of stay. For instance, the variability in length of stay across different diagnoses and age highlights the complexity and severity of the conditions being treated.

Outliers Detection and Processing. Outliers were identified in both the diagnosis and age features. These outliers represent valid but extreme cases. They can often be identified by other features such as the associated diagnosis and dependency, which were not documented in the EMR for short-stay hospitalizations. Additionally, these cases are influenced by other factors such as specific treatments and medical practices, which are beyond the scope of the current analysis.

To ensure robustness and improve the reliability of our findings, we proceeded to remove these outliers. This was achieved by calculating the first (Q1) and third (Q3) quartile for the relevant features and excluding values outside the bounds defined as (Lower Bound, Upper Bound).

This preprocessing ensured that our analysis focused on the most reliable and representative data, minimizing the potential for skewness while maintaining the integrity of the dataset.

4.2 Modeling

In Sect. 2, we reviewed various machine learning algorithms commonly employed in similar analyses, ranging from classification and regression techniques to neural networks. Since our task involves predicting a continuous variable, we selected four regression models: three gradient boosting models-XGBoost, CatBoost (which efficiently handles categorical features by reducing preprocessing steps [23]), and LightGBM (a gradient boosting model optimized for speed and efficiency)-as well as an ensemble learning model, Random Forest. We also applied a Feedforward Neural Network model (FNN), a widely used model where the information follows a unique direction.

These models were chosen for their established effectiveness in regression tasks and their capability to capture complex relationships in the data. Given that our dataset includes both categorical and continuous features, we applied label encoding to the categorical features, such as ICD-10 diagnosis codes and gender, for all models except CatBoost.

Model Hyperparameter Tuning. Hyperparameters are parameters that must be initialized before training and play a crucial role in optimizing model performance, which ultimately leads to better results [24,25]. In our study, we adopted the grid search approach [24]. This method evaluates all possible combinations of hyperparameters to identify the optimal configuration for achieving the best performance. Table 3 summarizes the hyperparameters used for each model.

Additionally, in the case of the FeedForward Neural Network, we used a four-layer architecture consisting of an input layer, two hidden layers, and an output layer. The input and hidden layers utilized the Rectified Linear Unit (ReLU) activation function to capture non-linear patterns, while the output layer adopted a linear activation function appropriate for predicting a continuous target variable.

Table 3. Optimal Hyperparameters of our models.

Model	Hyperparameters	Best Values
XGBoost	learning_rate, max_depth, n_estimators, subsample	0.01, 5, 500, 0.8
RF	min_sample_leaf, max_depth, n_estimators, min_samples_split	4, 10, 500, 10
CatBoost	depth, iterations, l2_leaf_reg, learning_rate	4, 500, 5, 0.1
LightGBM	learning_rate, n_estimators, num_leaves, subsample	0.01, 500, 31, 0.7
FNN	number of layers, activation functions, number of units	4, {ReLu, linear}, {64, 32, 1}

As shown in Table 3, four key hyperparameters were selected for each model and tested to identify the optimal configuration. These parameters were utilized during the training and validation phases, which will be further discussed in Sect. 4.3.

4.3 Evaluation and Results

The models we employed followed a conventional process, with the dataset split into training and test subsets, accounting for 80% and 20% of the data, respectively.

To evaluate the performance of our models, we relied on a set of well-established metrics commonly used in regression tasks. These metrics-Mean Absolute Error (MAE), Root Mean Squared Error (RMSE), and R^2 Score-allowed us to comprehensively assess the models' performance. Specifically, MAE measured the average magnitude of errors regardless of direction (positive or negative), RMSE quantified the margin between predictions and actual outcomes, and R^2 Score indicated the proportion of variance explained by the model, providing a measure of prediction precision [6]. Table 4, represents all the results of our models.

Based on the results presented in Table 4, the precision of the models is consistent, with minimal differences between the R^2 scores for the training and test datasets. This indicates no significant overfitting, as the models maintain similar performance levels on unseen data.

Table 4. Performance metrics for predicting Length of Stay.

Model	Train			Test		
	MAE	RMSE	R^2	MAE	RMSE	R^2
XGBoost	0.609	1.008	0.79	0.627	1.080	0.76
Random Forest	0.585	0.974	0.81	0.623	1.090	0.76
CatBoost	0.614	1.006	0.80	0.633	1.083	0.76
LightGBM	0.60	1.07	0.80	0.60	1.07	0.77
FeedForward NN	0.6	1.00	0.8	0.7	1.4	0.73

The LightGBM model yielded the best performance from our experiments, closely followed by XGBoost. Despite being specifically designed to handle categorical data effectively, CatBoost was outperformed by LightGBM. This can be attributed to LightGBM's robust hyperparameter tuning, which effectively mitigates overfitting, its efficient processing of categorical features, and its architecture, which allows it to explore information deeply and comprehensively.

However, the FeedForward Neural Network (FNN) was the least effective model, exhibiting the highest test RMSE among all models. However, the small difference between the train and test scores suggests that the model was not overfitting. This can be credited to early stopping during training, which prevented the model from excessively optimizing the training data.

Going further we explored the best model results to explain the importance and the influence of the variables using Shapley Additive exPlanations (SHAP) [26].

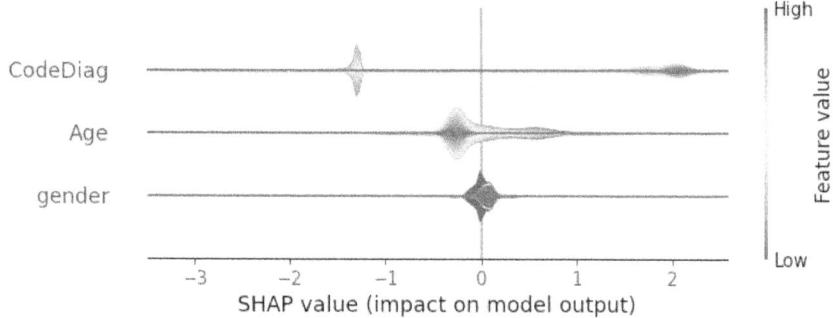

Fig. 2. Distribution of SHAP values across the features.

Figure 2, illustrates the importance and impact of each feature on the model's results. It shows that the diagnoses are the most influential factor, with two distinct groups: one group influences the SHAP positively, explained by a longer length of stay, and the other group influences the SHAP negatively, associated with a shorter length of stay. Next in importance is age, where older patients tend to have higher SHAP values. Finally, gender does not appear to significantly influence the model's predictions.

Additionally, as shown in Fig. 3, there is a non-linear relationship between age and LOS, which the model has effectively captured. We can also confirm the impact of these features and observe how age influences the diagnoses. Specifically, for patients aged 15 to 40, the SHAP values remain stable across a mix of diagnoses. However, for patients over 40, the SHAP values increase with age, following a positive trend. Finally, the interaction with diagnoses further supports this finding.

In conclusion, the interaction between age and diagnoses influences the SHAP values, with changes in age affecting the length of stay for different diagnoses. As age increases, it impacts the length of stay.

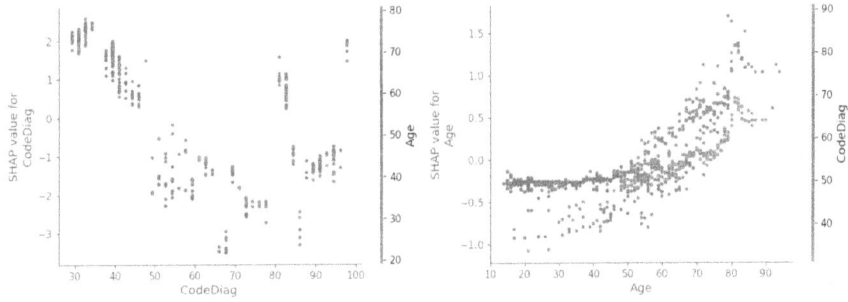

Fig. 3. SHAP Value Analysis for Diagnosis and Age.

5 Conclusion

In this study, we utilized machine learning algorithms to predict patients' surgical ward Length of Stay (LOS) at the time of admission, providing valuable insights for short-stay hospitalization. We analyzed electronic Medical records (EMR) comprising data from 7880 surgical ward patients after preprocessing. Our models included XGBoost, CatBoost, LightGBM, Random Forest, and a Feedforward Neural Network, achieving comparable performance with R^2 scores ranging from 73% to 77%, with LightGBM delivering the best performance of 77% R^2 and 0.6 MAE.

These results demonstrate that our approach outperforms related studies in LOS prediction at admission despite relying on a more limited feature set. Unlike other studies that incorporate both demographic and comprehensive clinical data available up to discharge, our analysis leveraged only demographic attributes (age and gender), as well as the principal diagnosis at admission.

Despite these promising results, certain limitations must be acknowledged. First, the model's reliance on limited data at admission may exclude valuable temporal patterns and longitudinal information in post-admission data. Additionally, the absence of key admission-time details such as secondary diagnoses, comorbidities, and the reason for admission, which are typically recorded at admission but unavailable in our dataset may have restricted the model's ability to capture important predictive factors. Furthermore, the generalizability of the findings is constrained by the dataset's single-center origin, which might limit applicability in diverse clinical settings. Expanding the dataset to include multiple hospital wards and a broader range of medical conditions would improve its robustness and applicability. Moreover, leveraging currently underutilized admission-time features -previously unusable due to missing values- could enhance predictive accuracy.

Looking ahead, several avenues for future research could strengthen and extend our work. In addition to dataset expansion, integrating electronic health records (EHRs) and real-time patient monitoring data could further enrich the predictive model by capturing evolving patient conditions.

We also aim to develop near real-time LOS and treatment predictions, potentially utilizing machine learning models such as recurrent neural networks (RNNs) or transformer-based architectures to enable dynamic adjustments in nurses' workload forecasts. By incorporating a more comprehensive set of features and leveraging temporal data more effectively, future models could better capture variations throughout a patient's hospital stay, ultimately improving resource allocation and operational efficiency in healthcare settings.

References

1. World Health Organization: How can hospital performance be measured and monitored?. In: How Can Hospital Performance be Measured and Monitored?, p. 17 (2003)
2. Tipton, K., Leas, B.F., Mull, N.K., et al.: Interventions to decrease hospital length of stay [internet]. Agency for Healthcare Research and Quality (US), Rockville (2021). (Technical Brief, No. 40) Introduction. https://ncbi.nlm.nih.gov/books/NBK574438/
3. Hunter, A., Johnson, L., Coustasse, A.: Reduction of intensive care unit length of stay: the case of early mobilization. Health Prog. **33**(2), 128–135 (2014)
4. Wang, Z., Chen, X., Wu, Y., et al.: A robust and interpretable ensemble machine learning model for predicting healthcare insurance fraud. Sci. Rep. **15**, 218 (2025). https://doi.org/10.1038/s41598-024-82062-x
5. Nadella, G., Satish, S., Meduri, K., Meduri, S.: A systematic literature review of advancements, challenges and future directions of AI and ML in healthcare. Int. J. Mach. Learn. Sustain. Dev. **5**, 115–130 (2023)

6. Almeida, G., Brito Correia, F., Borges, A.R., Bernardino, J.: Hospital length-of-stay prediction using machine learning algorithms-a literature review. Appl. Sci. **14**, 10523 (2024). https://doi.org/10.3390/app142210523
7. Erdogan Yildirim, A., Canayaz, M.: Machine learning-based prediction of length of stay (LoS) in the neonatal intensive care unit using ensemble methods. Neural Comput. Appl. **36**, 14433–14448 (2024). https://doi.org/10.1007/s00521-024-09831-7
8. Deschepper, M., De Smedt, C., Colpaert, K.: A literature-based approach to predict continuous hospital length of stay in adult acute care patients using admission variables: a single university center experience. Int. J. Med. Inform. **193** (2025). https://doi.org/10.1016/j.ijmedinf.2024.105678
9. Zhang, M., Kuo, T.-T.: Early prediction of long hospital stay for intensive care units readmission patients using medication information. Comput. Biol. Med. **174** (2024). https://doi.org/10.1016/j.compbiomed.2024.108451
10. Peres, I.T., Hamacher, S., Oliveira, F.L.C., Bozza, F.A., Salluh, J.I.F.: Data-driven methodology to predict the ICU length of stay: a multicentre study of 99,492 admissions in 109 Brazilian units. Anaesth. Crit. Care Pain Med. **41** (2022). https://doi.org/10.1016/j.accpm.2022.101142
11. Alsinglawi, B., Alshari, O., Alorjani, M., et al.: An explainable machine learning framework for lung cancer hospital length of stay prediction. Sci. Rep. **12**, 607 (2022). https://doi.org/10.1038/s41598-021-04608-7
12. Chrusciel, J., Girardon, F., Roquette, L., et al.: The prediction of hospital length of stay using unstructured data. BMC Med. Inform. Decis. Mak. **21**, 351 (2021). https://doi.org/10.1186/s12911-021-01722-4
13. Lee, H., Kim, S., Moon, H., et al.: Hospital length of stay prediction for planned admissions using observational medical outcomes partnership common data model: retrospective study. J. Med. Internet Res. **26**, e59260 (2024). https://www.jmir.org/2024/1/e59260, https://doi.org/10.2196/59260
14. Grampurohit, S., Sunkad, S.: Hospital length of stay prediction using regression models. In: 2020 IEEE International Conference for Innovation in Technology (INOCON), pp. 1–5. IEEE Press, Bengaluru (2020). https://doi.org/10.1109/INOCON50539.2020.9298294
15. Abdurrab, I., Mahmood, T., Sheikh, S., et al.: Predicting the length of stay of cardiac patients based on pre-operative variables-bayesian models vs. machine learning models. Healthc. (Basel) **12**(2), 249 (2024). https://doi.org/10.3390/healthcare12020249
16. Harerimana, G., Kim, J.W., Jang, B.: A deep attention model to forecast the length of stay and the in-hospital mortality right on admission from ICD codes and demographic data. J. Biomed. Inform. (2021). https://doi.org/10.1016/j.jbi.2021.103778
17. Johnson, A., Pollard, T., Mark, R.: MIMIC-III clinical database demo (version 1.4). PhysioNet (2019). https://doi.org/10.13026/C2HM2Q
18. Futura Smart Design. https://www.calystene.com/futura-smart-design-solutions-informatiques-pour-hopitaux-et-cliniques/. Accessed 19 Dec 2024
19. Gharbi, M., Verdier, C., Di Mascolo, M.: Nurses' workload prediction in hospitals-a machine learning-based approach. In: INFORSID (2024)
20. Mirza, B., Li, X., Lauwers, K., et al.: A clinical site workload prediction model with machine learning lifecycle. Healthc. Anal. **3**, 100159 (2023)
21. ICD-10. https://icd.who.int/browse10/2019/en. Accessed 02 Jan 2025
22. Mean LOS. https://www.atih.sante.fr/sites/default/files/public/content/4729/synthese_aah_2022_mco.pdf. Accessed 10 Jan 2025

23. Prokhorenkova, L., Gusev, G., Vorobev, A., Dorogush, A.V., Gulin, A.: CatBoost: unbiased boosting with categorical features. In: NIPS 2018: Proceedings of the 32nd International Conference on Neural Information Processing Systems, Montreal, pp. 6639–6649 (2018)
24. Ogunsanya, M., Isichei, J., Desai, S.: Grid search hyperparameter tuning in additive manufacturing processes. Manuf. Lett. **35**(Suppl.), 1031–1042 (2023). https://doi.org/10.1016/j.mfglet.2023.08.056
25. Belete, D., Manjaiah, D.H.: Grid search in hyperparameter optimization of machine learning models for prediction of HIV/AIDS test results. Int. J. Comput. Appl. **44**, 1–12 (2021). https://doi.org/10.1080/1206212X.2021.1974663
26. Lundberg, S.: A unified approach to interpreting model predictions. arXiv preprint arXiv:1705.07874 (2017)

Alignment of Schema-Only and Instance-Only Data Sources Using Large Language Models

Nour Elhouda Kired[1,2](✉), Franck Ravat[1], Jiefu Song[1], and Olivier Teste[2]

[1] IRIT - Université Toulouse Capitole, Toulouse, France
{nour-elhouda.kired,franck.ravat,jiefu.song}@irit.fr
[2] IRIT - Université Toulouse II Jean Jaurès, Toulouse, France
olivier.teste@irit.fr

Abstract. Data matching is a persistent challenge in heterogeneous data integration. Traditional data matching methods, relying on schema-based, instance-based, or hybrid approaches, often fall short when aligning disparate data from schema-only with instance-only sources. To address this problem of aligning disparate sources, we present an innovative data matching framework that enables the matching of one data source, where only schema-related information is available, with another data source, where only instance-related information is available. The strength of our framework lies in its ability to combine outputs from multiple Schema-Instances matchers, generating auxiliary information to enhance the alignment between disparate data structures. Our framework is validated using the Valentine Benchmark through extensive experiments. These findings underscore the potential of our approach to advance the integration of diverse data sources.

Keywords: Data matching · Disparate Data Source Structures · Auxiliary Information Generation · Matcher Combination

1 Introduction

Data lakes have emerged as versatile platforms for integrating diverse data types, including structured, semi-structured, and unstructured data, within modern information systems. These platforms enable organizations to manage heterogeneous data sources at scale. However, data matching, a foundational step in aligning data for integration, faces significant challenges when working with schema-only or instance-only sources, often arising due to privacy concerns or incomplete datasets [1]. While traditional data matching [2] techniques-such as schema-based matching [3], instance-based matching [4], and hybrid methods [5]-have proven effective in structured contexts, their performance diminishes when dealing with disparate data structures or missing instances. Addressing these limitations is crucial for advancing the seamless integration capabilities of data-driven systems.

A further complexity arises with instance-only data sources, where the schema is either missing or incomplete, making it challenging to match them with schema-only sources. While established methods exist for aligning data that includes both schema and instances, there is a significant gap in addressing the alignment between schema-only with instance-only data. Our previous research [6] has explored the alignment of schema-only sources with datasets that include both schema and instance data using cross-domain and domain-specific embeddings. This is where the need for advanced techniques, capable of inferring auxiliary informations, becomes critical [7–10].

This paper addresses the alignment of schema-only and instance-only structured data sources by leveraging large language models (LLMs) to generate synthetic schemas and instances. It combines simple and hybrid matchers to align these enriched data sources. Our contributions include a generic schema-instance matching framework, implemented with two LLMs and five basic matchers, and validated through comprehensive experiments demonstrating its effectiveness, efficiency, and robustness. The structure of this paper is as follows: Section. 2 reviews the existing literature on data matching. Section 3 highlights our contributions, introduces the proposed framework, and details the methodology. Section 4 presents a comprehensive evaluation of our approach. Finally, Section. 5 concludes and discusses directions for future research.

2 Related Work

Data matching is a fundamental task in data integration and knowledge discovery, particularly when working with heterogeneous data sources. Over the years, researchers have developed numerous solutions that vary depending on the nature of the data and the specific challenges posed by different data formats. Broadly, data matching approaches are categorized as schema-based, instance-based, and hybrid techniques [1]. Schema-based matchers use structural elements (attribute names, data types, descriptions) to identify correspondences, while instance-based matchers leverage patterns or statistical distributions from actual data. Hybrid matchers combine both approaches to enhance matching accuracy, as in COMA++ [5], which integrates multiple algorithms and external resources like WordNet to resolve synonyms and provide semantic context. These matchers operate at different levels of granularity, including element-level (focusing on individual attributes) and structure-level (comparing higher-order structures like tables or classes).

Recent advancements in schema discovery [7] have presented solutions for inferring schemas from both structured and semi-structured data. These methods are particularly valuable in scenarios where schema information is incomplete or entirely absent, as seen with instance-only datasets. In such cases, the ability to deduce semantic relationships or constraints directly from the data itself becomes a critical tool for addressing schema alignment challenges.

On the other hand, methods for generating synthetic instances from existing schemas tackle the inverse problem. For instance, the approach described in [8,10] provides a framework for creating synthetic data that adheres to a given schema. This capability is essential for checking schema relationships. Together, these methods complement schema inference techniques by ensuring that both schema alignment and instance creation needs are addressed [7].

Furthermore, linguistic resources such as WordNet and large language models (LLMs), such as GPT-3.5 [11] and Llama-3.2 [12], have introduced new possibilities for improving both schema and instance inference. WordNet, a lexical database, resolves synonymy between terms, enhancing schema alignment when attribute names vary across datasets [13]. Generative LLMs, however, offer a unique advantage with their ability to simultaneously perform schema inference and instance generation. By interpreting both structural and semantic aspects of data, they bridge gaps between schema-only and instance-only sources, eliminating the need for separate tools to handle these tasks [14]. Unlike traditional methods that depend on predefined rules or templates, LLMs dynamically adapt to diverse data domains and contexts. Leveraging their vast pretrained knowledge, they generate auxiliary information tailored to specific attributes, data types, and domain semantics [10].

Traditional matching methods-such as COMA++ [5], Similarity Flooding, or string similarity approaches like Jaccard and Levenshtein-can be applied when synthetically generated schemas and instances are available. However, these methods struggle to handle the noise and inconsistencies introduced by synthetic data, which can be vague or significantly different from the original schema. Probabilistic frameworks like PARIS [15], which jointly align entities, classes, and relations through probabilistic inference, also rely on co-occurrence statistics and distributional assumptions. In the presence of noisy, incomplete, or highly heterogeneous data, these assumptions often do not hold. As a result, such probabilistic methods may not be effective in our scenario, and we have not validated their performance in this context. Instead, we leverage LLMs as matchers because of their ability to abstract from generation noise and to capture deep semantic and syntactic correspondences. This makes them more robust when dealing with heterogeneous and noisy data sources [16]. Unlike traditional or rule-based approaches, LLMs offer greater adaptability to variations in data structure, terminology, and representation.

Building on these advancements, our proposed framework integrates generative AI models GPT-3.5 [11] and Llama-3.2 [12] with linguistic and constraint-based techniques to deliver a comprehensive solution for aligning disparate data sources. By addressing both schema and instance-level challenges, the framework ensures robust data matching and synthesis, even when information is incomplete. This unified approach simplifies the complexities of heterogeneous data integration and sets a new standard for handling disparate data environments.

3 Proposed Framework

3.1 Overview of the Framework

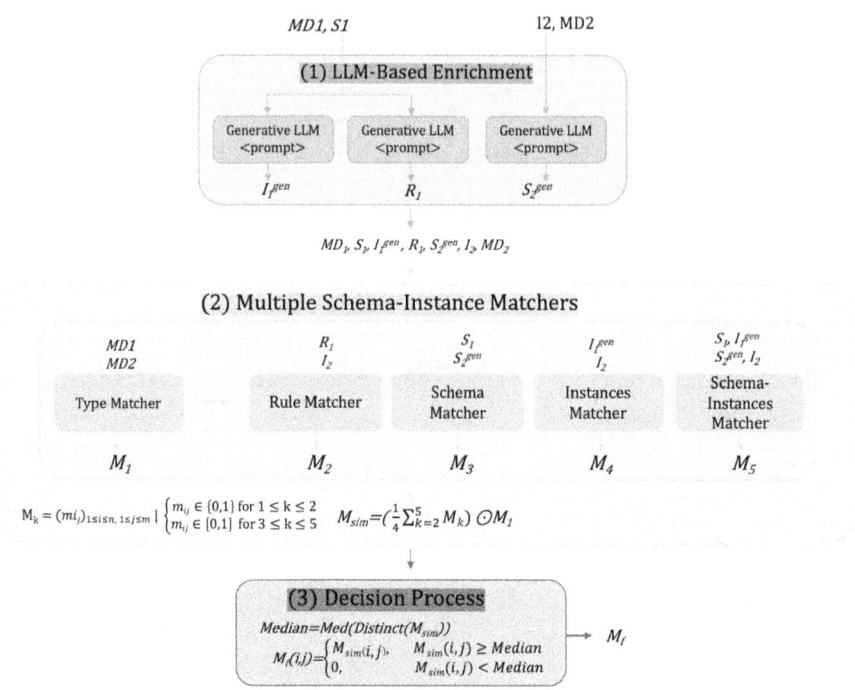

Fig. 1. Overview of the Proposed Framework

Our proposed framework, illustrated in Fig. 1, addresses the alignment of schema-only (S_1) and instance-only (I_2) data sources through the following component. Table 1 provides a legend and annotations for clarity.

1. **LLM-Based Enrichement:** uses large language models (GPT-3.5 or Llama-3.2) to generate auxiliary information, such as synthetic instances and constraint rules (I_1^{gen} and R_1) for S_1, generated schemas (S_2^{gen}) for I_2. This enrichment provides the semantic and structural context needed for effective alignment.
2. **Basic Schema and Instance Matchers:** Pre-matchers are applied to compare the two data sources and eliminate data that cannot be matched (e.g., incompatible types like string and number). In this paper, we use the Type Matcher as a pre-matcher for this initial comparison, which checks data types and formats from metadata, the Rule Matcher validates alignment using constraints (e.g., ranges, regex), the Schema Matcher aligns structural elements, the Instances Matcher compares instance values, and the Schema-Instances

Table 1. Legend and Annotations for the Framework

Legend	Annotation
S_1	$S_1 = \{a^1, a^2, \ldots, a^n\}$: Schema of Data Source 1, where each a^i represents an attribute.
I_1^{gen}	$I_1^{\text{gen}} = \{V^1, \ldots, V^n\}$: Set of possible values for each attribute a^i in S_1. Each $V^i = \{v_1^i, \ldots, v_{p_i}^i\}$ represents the possible values for the attribute a^i, $\forall i \in [1, n]$.
S_2^{gen}	$S_2^{\text{gen}} = \{S_2^1, \ldots, S_2^m\}$: Schemas derived from Data Source 2, where each $S_2^j = \{b_1^j, \ldots, b_{K_j}^j\}$ lists the attributes generated at the j-th position, $\forall j \in [1, m]$.
I_2	$I_2 = \{W^1, \ldots, W^m\}$: Set of values for each S_2^j in S_2^{gen}. Each $W^j = \{w_1^j, \ldots, w_{q_j}^j\}$ denotes the values for the attributes in S_2^j, $\forall j \in [1, m]$.
MD_1	Metadata associated with S_1, including types for each attribute a^i: $Types_1 = \{types(a^i) \mid i \in \{1, \ldots, n\}\}$, and $name_{S_1}$, which defines the name of Data Source 1.
MD_2	Metadata associated with I_2, including types for each W^j: $Types_2 = \{types(W^j) \mid j \in \{1, \ldots, m\}\}$, and $name_{I_2}$, which defines the name of Data Source 2.
R_1	Rules associated with S_1, enforcing constraints on attributes a^i: regular expressions $W^j \in$ regex(pattern), string length $l_{\min} \leq \text{len}(W^j) \leq l_{\max}$, range $v_{\min} \leq W^j \leq v_{\max}$, and uniqueness count$(W^j) = 1, \forall j \in [1, m]$.
M_k	Result matrix for the matchers: Type Matcher (1), Rule Matcher (2), Schema Matcher (3), Instances Matcher (4), and Schema-Instances Matcher (5). Each $m_{k,ij}$ represents the similarity score between a^i (from S_1) and W^j (from I_2): $$M_k = (m_{ij})_{1 \leq i \leq n,\ 1 \leq j \leq m} \quad \Big| \quad \begin{cases} m_{ij} \in \{0, 1\}, & \text{if } k \in \{1, 2\} \\ m_{ij} \in [0, 1], & \text{if } k \in \{3, 4, 5\} \end{cases}$$
M_{sim}	Similarity matrix, where $m_{\text{sim},ij}$ represents the similarity score between a^i (from S_1) and W^j (from I_2): $$M_{\text{sim}} = (m_{\text{sim},ij})_{1 \leq i \leq n,\ 1 \leq j \leq m}.$$

matcher combines schema and instance features for hybrid alignment. Each matcher produces a similarity matrix representing the confidence of alignment.

3. **Decision Process:** aggregates the similarity matrices into a unified matrix (M_{sim}). Thresholding using median filtering is then applied to produce the final matched attribute pairs (M_f) and their corresponding similarity scores. The final output consists of matched attribute pairs with corresponding matching scores that represent the alignment between schema-only and instance-only data.

3.2 Detailed Framework Components

1. LLM-Based Enrichment. Forms the foundation of our framework by leveraging large language models (LLMs), such as GPT-3.5 or Llama-3.2, to enrich schema-only (S_1) and instance-only (I_2) data sources. This phase generates auxiliary information to address data representation gaps, enabling effective alignment. Below, we detail the key operations and the prompts used to generate the enriched outputs.

1.1. Generating Synthetic Instances for S_1. To create synthetic instances for each attribute in the schema-only data source (S_1), the following prompt is used:

> *Provide a list of possible values that are meaningful and representative of the given attribute a^i and its type types(a^i), within the context of name$_{S_1}$. The output should be strictly in the format of a Python list, e.g.,* `[value1, value2, ...]`. *Return only the list as the output, without any additional comments or explanations.*

This ensures that synthetic instances generated for S_1 accurately reflect the semantics of the attribute and its domain. The output is used in Schema-Instances and Instances matchers.

1.2. Generating Missing Attributes for I_2. To construct a plausible schema for the instance-only data source (I_2), we derive potential attribute names based on the provided instance values using the following prompt:

> *Provide a list of possible attribute names that are meaningful and representative of the given values W^j and type types(W^j) in the context of the name$_{I_2}$. The output should strictly be in the format of a Python list, e.g., [attribute1, attribute2, ...]. Only return the list as the output, without additional comments or explanations.*

This step generates synthetic schemas S_2^{gen} for I_2, ensuring that the instance data has a structural representation compatible with S_1. The output is used in Schema-instances and Schema Matchers.

1.3. Rules and Constraints Generation. To further support alignment, strict validation rules and constraints are generated for each attribute in S_1. These rules capture important properties such as data types, ranges, formats, and uniqueness. We designed the following prompt:

> *Generate strict validation rules for the attribute a^i in the dataset name$_{S_1}$ with the data type types(a^i). Return the rules as a Python list, with each rule represented as a string. Focus on constraints such as ranges, formats using regular expressions, length checks, and logical consistency.*

The output includes validation rules (R_1) defining acceptable values. For instance, given a schema S containing an attribute age, a possible rule would be: type(age) = integer and age $\in [0, 120]$.

These outputs bridge the semantic gaps through enrichment using LLMs and address structural issues by generating missing elements from both data sources to resolve inconsistencies between S_1 and I_2, providing a robust foundation for subsequent matchers.

2. Multiple Schema-Instances matchers. This component compares and aligns attributes and instances between schema-only data (S_1) and instance-only data (I_2). It employs multiple matchers, detailed below.

2.1. Type Matcher (M_1) validates compatibility between attributes in S_1 and instances in I_2 based on their data types:

$$M_1 = (m_{1,ij})_{1 \leq i \leq n,\ 1 \leq j \leq m}, \quad m_{1,ij} = \mathbb{1}_{\{\texttt{types}(a^i) = \texttt{types}(W^j)\}}, \quad m_{1,ij} \in \{0, 1\}.$$

Attributes with mismatched types **are filtered out** to avoid wasting resources.

2.2. Rule Matcher (M_2) checks if instance values in I_2 satisfy predefined rules R_1 in S_1:

$$M_2 = (m_{2,ij})_{1 \leq i \leq n,\ 1 \leq j \leq m}, \quad m_{2,ij} = \mathbb{1}_{\{W^j \text{ satisfies all } R_{1_i}\}}, \quad m_{2,ij} \in \{0, 1\}.$$

2.3. Schema Matcher (M_3) computes semantic similarity between S_1 and S_2^{gen} using embeddings:

$$M_3 = (m_{3,ij})_{1 \leq i \leq n,\ 1 \leq j \leq m}, \quad m_{3,ij} = \frac{a^i \cdot S_2^j}{\|a^i\| \|S_2^j\|}, \quad m_{3,ij} \in [0, 1].$$

2.4. Instances Matcher (M_4) compares I_1^{gen} and I_2 using cosine similarity:

$$M_4 = (m_{4,ij})_{1 \leq i \leq n,\ 1 \leq j \leq m}, \quad m_{4,ij} = \frac{V^i \cdot W^j}{\|V^i\| \|W^j\|}, \quad m_{4,ij} \in [0, 1].$$

2.5. Schema-Instances Matcher (M_5), a hybrid matcher that combines schema and instance representations into unified embeddings:

$$M_5 = (m_{5,ij})_{1 \leq i \leq n,\ 1 \leq j \leq m}, \quad m_{5,ij} = \frac{\boldsymbol{DS_1^i} \cdot \boldsymbol{DS_2^j}}{\|\boldsymbol{DS_1^i}\| \|\boldsymbol{DS_2^j}\|}, \quad m_{5,ij} \in [0, 1].$$

where:

- $\boldsymbol{DS_1^i}$ = concat(a^i, V^i): The embedding of the concatenated i-th schema attribute (a^i) and its corresponding instance information (V^i).
- $\boldsymbol{DS_2^j}$ = concat(S_2^j, W^j): The embedding of the concatenated j-th schema attribute (S_2^j) and its corresponding instance information (W^j).

Embedding generation was applied in the Schema Matcher, Instances Matcher, and Schema-Instances Matcher, using large language models (LLMs). These models included BERT, RoBERTa, DistilBERT, ALBERT, Bart, the Sentence-transformer model `all-MiniLM-L6-v2`, and OpenAI's GPT-3.5.

3. Integration and Aggregation. The output from all matchers is integrated into a final similarity matrix M_{sim}, defined as:

$$M_{\text{sim}} = (m_{\text{sim},ij})_{1 \leq i \leq n,\ 1 \leq j \leq m}, \quad m_{\text{sim},ij} = \left(\frac{1}{4} \sum_{k=2}^{5} M_k \right) \odot M_1, \quad m_{\text{sim},ij} \in [0, 1].$$

where $\odot$ represents the Hadamard product, which ensures the final similarity retains the type compatibility constraints.

Decision Process. The decision process refines the similarity scores in the matrix M_{sim} to generate the final similarity matrix M_f. This matrix is used to decide whether each pair is accepted or rejected, based on a threshold calculated from the median of the distinct similarity scores. The process is defined as follows:

Median of Distinct Similarity Scores. The median is computed from the unique values in M_{sim}:

$$\text{Median} = \text{Med}(\text{Distinct}(M_{\text{sim}})).$$

The *Distinct* function extracts a list of unique values from the matrix M_{sim}, and the *Median* function calculates the central value of this list. This ensures that the threshold reflects the distribution of similarity scores while mitigating the impact of repeated low or high values.

Apply Thresholding. Each entry in the similarity matrix $M_{\text{sim}}[i,j]$ is compared against the median threshold. The decision process produces a refined similarity matrix M_f, defined as:

$$M_f = (m_{f,ij})_{1 \leq i \leq n,\ 1 \leq j \leq m}, \quad m_{f,ij} = \begin{cases} m_{\text{sim}}[i,j], & \text{if } m_{\text{sim}}[i,j] \geq \text{Median}, \\ 0, & \text{otherwise}. \end{cases}, \quad m_{f,ij} \in [0,1].$$

This step ensures that only pairs with similarity scores above the median are retained, while the rest are rejected by setting their similarity scores to zero.

Final Output. The resulting decision matrix M_f is constrained to the range $M_f \in \mathbb{R}^{n \times m}$, where n and m represent the number of attributes in S_1 and instances in I_2, respectively. The nonzero entries in M_f represent the accepted schema-instances alignments with their corresponding similarity scores. The decision process serves as the final step in the framework, transforming the aggregated similarity matrix into actionable results for schema-instances matching.

4 Experiments

In this section, we present our experiments aimed at evaluating our data matching approach. We formulate our investigations around the following research questions:

- **RQ1 (Effectiveness)** Does the framework perform effectively across various datasets in terms of F1 score and NDCG? Which combination of large language models (LLM) -one for generating auxiliary information and another for generating embeddings- yields the best performance in these metrics?
- **RQ2 (Baseline)** How does the proposed framework compare to traditional data matching methods-such as COMA++, Similarity Flooding, or Jaccard/Levenshtein-in terms of effectiveness (F1 Score) across all datasets?
- **RQ3 (Efficiency)** Which large language model (LLM) provides the best performance in terms of execution time across various datasets?

- **RQ4 (Noise Sensitivity)** How effectively does the proposed data matching solution perform across varying levels of dataset noise? How consistent is it in maintaining F1 Score and NDCG?
- **RQ5 (Ablation Study)** How does the ablation of our framework matchers impact the performance in terms of F1 Score and NDCG across diverse datasets?

4.1 Datasets

Table 2 presents four diverse datasets from the Valentine Benchmark [17] - *TPC-DI*, *ChEMBL*, *WikiData*, and *Magellan Data*-designed to vary in attribute overlap, syntactic noise, and structural relationships. These datasets simulate real-world data integration challenges.

TPC-DI represents an Online Transaction Processing (OLTP) system, characterized by moderately structured data. *ChEMBL* is a bioinformatics dataset containing bioactive molecules with drug-like properties, offering a unique domain-specific challenge due to complex attribute naming conventions. *WikiData* focuses on music-related datasets, emphasizing structural relationships between schemas. Finally, the *Magellan Data* suite comprises several widely-used dataset pairs, such as Amazon-Google, Walmart-Amazon, and IMDB-Movielens, with controlled conditions to test the robustness of data matching techniques.

Table 2. Dataset Characteristics

Dataset Source (#Pairs)	#Rows	#Attributes	#Matches
TPC-DI (180)	7.5k - 15k	11 - 22	1, 6, 11, 15, 22
ChEMBL (180)	7.5k - 15k	12 - 23	1, 6, 11, 16, 23
WikiData (4)	5.4k - 10.8k	13 - 20	6, 8, 20
Magellan (7)	331 - 66.8k	4 - 9	4 - 9

In our experiments, we adopted a controlled setup to rigorously evaluate the data matching framework. Specifically, for each dataset pair, we used the schema from one dataset as the **schema-only source** and the instances from the second dataset as the **instance-only source**. This approach ensures a clear separation of schema and instance data, reflecting real-world scenarios where complete information from both sources may not always be available. This setup allows us to test the framework's robustness in aligning disparate data sources with minimal prior information.

4.2 Experimental Setup

We deploy our approach and its variations using PyTorch version 1.6.0 with CPU support. The experiments were conducted on a system equipped with an Intel

Core i7-10750H processor, featuring 6 cores running at 2.6 GHz (up to 5.0 GHz with Turbo Boost). The system has 16 GB of RAM, which provides sufficient memory for the experiments.

The framework was evaluated by testing various models for generating auxiliary information and embeddings. For generating auxiliary information, two models, GPT-3.5 and Llama-3.2, were used. Additionally, seven different embedding models were tested across schema, instance, and hybrid matchers, including BERT(bert-base-uncased), RoBERTa, DistilBERT(distilbert-base-uncased), ALBERT (albert-base-v2), Bart(facebook/bart-base), the Sentence-transformer model (sentence-transformers/all-MiniLM-L6-v2), and GPT-3.5. The performance of these combinations was assessed using the F1 score and NDCG metrics. The Type and Rule matchers stayed fixed because they check types and rules, which give binary results. For further details, please refer to the companion website[1].

4.3 RQ1. Effectiveness

Table 3 highlights that the framework achieves modest results across all combinations, with consistent performance observed for both the F1 score and NDCG. On average, GPT-3.5 achieves an F1 score of 0.53 and an NDCG of 0.71, while Llama-3.2 yields an average F1 score of 0.48 and an NDCG of 0.62. These results suggest the framework works effectively, with GPT-3.5 showing a slight edge over Llama-3.2.

Table 3. Performance of LLMs for auxiliary information generation and embedding tasks based on F1 Score and NDCG (mean ± std)

Model	F1 Score		NDCG	
	GPT-3.5	Llama-3.2	GPT-3.5	Llama-3.2
ALBERT	**.52** ± .29	.48 ± .25	**.66** ± .30	.51 ± .29
BERT	**.54** ± .29	.47 ± .25	**.70** ± .29	.64 ± .29
DistilBERT	**.54** ± .28	.46 ± .26	**.71** ± .27	.64 ± .27
BART	**.52** ± .29	.42 ± .23	**.70** ± .29	.54 ± .30
GPT	**.58** ± .26	.50 ± .25	**.84** ± .18	.79 ± .24
RoBERTa	**.53** ± .26	.43 ± .25	**.67** ± .29	.51 ± .24
MiniLM	**.54** ± .27	.45 ± .25	**.71** ± .30	.68 ± .26

---

[1] https://github.com/user28060/Alignment-of-schema-only-and-instance-only-data.git.

Table 4. Wilcoxon signed-rank test results (GPT-3.5 vs Other models)

Model	Statistic	p-value
ALBERT	8276.0	9.60×10^{-18}
BERT	9386.5	6.30×10^{-9}
DistilBERT	10582.0	1.12×10^{-6}
BART	8595.0	3.10×10^{-16}
RoBERTa	6374.0	1.57×10^{-13}
MiniLM-L6-v2	9091.5	5.25×10^{-9}

Among embedding models, GPT embeddings achieve the best performance (with GPT-3.5 for generation), with an average F1 score of 0.58 and NDCG of 0.84, showing lower variability. Other models like DistilBERT (F1 = 0.54, NDCG = 0.71) and MiniLM (F1 = 0.54, NDCG = 0.71) perform slightly lower, while RoBERTa (F1 = 0.53, NDCG = 0.67) and BART (F1 = 0.52, NDCG = 0.70) show less effective results. To ensure the validity of these results, we applied the Wilcoxon signed-rank test—appropriate due to the non-normal distribution of our data [18]. The statistical analysis was performed in the context of data generated by GPT-3.5, where GPT embeddings were compared against other embedding models. The Wilcoxon test confirms that the superior performance of GPT embeddings over other models is statistically significant across all evaluations, with p-values consistently below 0.05 (see Table 4).

Beyond the performance of LLM models, the relationship type and the number of matches to identify within a pair of data sources could also be key factors. Datasets can be categorized into four types: *Unionable*, with compatible schemas for direct merging; *View-Unionable*, requiring minimal transformations; *Joinable*, linked through shared keys; and *Semantically-Joinable*, linked via semantic similarities. The *attribute overlap* percentage quantifies schema similarity, directly influencing data matching performance: high overlap enables easier merging, while low overlap necessitates complex techniques.

Table 5 shows that **Unionable relations** (100% overlap) lead to the best performance, with GPT embeddings achieving an F1 score of 0.83 and NDCG of 0.87. In contrast, more complex relations like **Semantically-Joinable** (8–82% overlap) result in a significant performance drop (mean F1 = 0.50). Figure 2 further confirms these observations, illustrating **that higher attribute overlap consistently leads to better results.**

Table 5. LLMs for auxiliary information generation and embedding tasks based on F1 Score and NDCG (mean ± std) across relations.

Model	Relation	F1 Score		NDCG	
		GPT-3.5	Llama-3.2	GPT-3.5	Llama-3.2
ALBERT	Joinable	**.44** ± .28	.41 ± .23	**.62** ± .32	.45 ± .29
	Semantically-Joinable	**.46** ± .28	.40 ± .24	**.65** ± .32	.47 ± .31
	Unionable	**.81** ± .10	.73 ± .12	**.83** ± .10	.71 ± .12
	View-Unionable	.40 ± .26	**.44** ± .21	**.56** ± .33	.45 ± .29
BERT	Joinable	**.46** ± .28	.40 ± .25	**.66** ± .32	.60 ± .32
	Semantically-Joinable	**.47** ± .28	.40 ± .26	**.68** ± .31	.55 ± .32
	Unionable	**.82** ± .11	.70 ± .11	**.84** ± .08	.79 ± .11
	View-Unionable	**.46** ± .26	.42 ± .22	**.66** ± .30	.63 ± .31
DistilBERT	Joinable	**.48** ± .27	.39 ± .24	**.67** ± .31	.60 ± .29
	Semantically-Joinable	**.47** ± .26	.40 ± .26	**.69** ± .28	.61 ± .28
	Unionable	**.82** ± .12	.71 ± .12	**.87** ± .08	.78 ± .10
	View-Unionable	**.45** ± .25	.41 ± .23	**.65** ± .28	.59 ± .29
BART	Joinable	**.46** ± .30	.35 ± .23	**.65** ± .34	.51 ± .33
	Semantically-Joinable	**.46** ± .28	.37 ± .24	**.70** ± .30	.51 ± .33
	Unionable	**.79** ± .10	.62 ± .09	**.82** ± .09	.73 ± .13
	View-Unionable	**.41** ± .28	.37 ± .23	**.65** ± .30	.46 ± .30
GPT	Joinable	**.51** ± .25	.42 ± .24	**.82** ± .23	.72 ± .31
	Semantically-Joinable	**.50** ± .25	.42 ± .24	**.82** ± .20	.72 ± .30
	Unionable	**.83** ± .11	.74 ± .13	**.87** ± .08	.86 ± .07
	View-Unionable	**.51** ± .25	.46 ± .22	**.87** ± .15	.87 ± .15
RoBERTa	Joinable	**.46** ± .25	.36 ± .24	**.65** ± .30	.50 ± .25
	Semantically-Joinable	**.47** ± .25	.37 ± .25	**.66** ± .31	.51 ± .26
	Unionable	**.78** ± .13	.69 ± .14	**.84** ± .09	.60 ± .16
	View-Unionable	**.45** ± .22	.37 ± .22	**.57** ± .31	.45 ± .22
MiniLM	Joinable	**.48** ± .27	.39 ± .24	**.69** ± .32	.65 ± .30
	Semantically-Joinable	**.47** ± .26	.38 ± .24	**.68** ± .34	.65 ± .30
	Unionable	**.80** ± .10	.68 ± .13	**.86** ± .08	.76 ± .10
	View-Unionable	**.47** ± .25	.38 ± .23	.63 ± .33	**.68** ± .25

Discussion. GPT-3.5 is the best choice for auxiliary information generation, producing high-quality inputs that enhance embedding performance. GPT embeddings consistently outperform other models across all tasks, delivering reliable

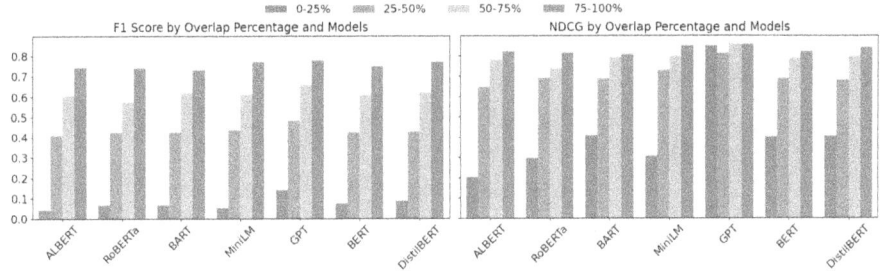

Fig. 2. Average F1 Score and NDCG for GPT-3.5 across embeddings models and overlap attributes.

Table 6. Baseline F1 and Top F1 Scores (Mean ± Standard Deviation)

Algorithme	Distance	F1 (mean ± std)	F1 (Top) (mean ± std)
COMA_INS	–	.33 ± .34	.33 ± .34
COMA_SCH	–	.33 ± .34	.33 ± .34
JaccardMatcher	DamerauLevenshtein	.30 ± .25	.31 ± .25
JaccardMatcher	Hamming	.31 ± .26	.31 ± .26
JaccardMatcher	JaroWinkler	.39 ± .25	.41 ± .25
JaccardMatcher	Levenshtein	.32 ± .26	.33 ± .26
SimilarityFlooding	–	.10 ± .11	.11 ± .12

and high-quality results with lower variability. While simpler Unionable relations achieve the best outcomes, the robustness of GPT embeddings and GPT-3.5's generation capabilities ensure competitive performance even in more complex relations requiring semantic reasoning.

4.4 RQ2. Baseline

To address RQ2, we compared the effectiveness of our LLM-based matching framework with traditional approaches, including COMA++, Similarity Flooding, and string similarity methods (Jaccard combined with Damerau-Levenshtein, Hamming, Jaro-Winkler, and Levenshtein). The experiments were conducted on 371 datasets. For each dataset, we generated auxiliary schemas and instances using GPT-3.5, sampling 100 schema variations for instances-only data sources and repeating the process 10 times, resulting in a total of 1000 executions per dataset.

In this setting, noise refers to the variability in the quality of the generated schemas. Some of the proposed schemas can significantly deviate from the real underlying schema, introducing inconsistencies and ambiguity in the matching process. In addition to the average F1 scores, we also report the Top F1 score. The Top F1 represents the best F1 score obtained for each dataset across all schema variations. It reflects the optimal potential performance of each method under ideal conditions, where the generated schema is most aligned with the target.

Discussion. Table 6 shows that traditional methods, such as COMA_INS and COMA_SCH, achieve low F1 scores (.33 ± .34), with minimal improvement in their Top F1 scores. Similarity Flooding performs worse, with an F1 of .10 ± .11. Among string similarity methods, Jaro-Winkler yields the highest F1 (.39 ± .25) and a Top F1 of .41 ± .25.

However, the limited gains between the average F1 and Top F1 scores suggest that traditional approaches struggle to adapt to heterogeneous and noisy data, even when the generated schemas are closer to the real ones. In contrast, our LLM-based framework significantly outperforms these baselines, offering higher accuracy and robustness across all datasets (see Table 3).

4.5 RQ3. Effeciency

Table 7. Execution times for auxiliary information generation models GPT-3.5 and Llama-3.2 across dataset source datasets and embeddings models.

(a) Execution times by Dataset Source.

Dataset Source	GPT (s)	Llama (s)
ChEMBL	212.89	128.34
TPC-DI	139.84	144.89
Wikidata	704.90	930.10
Magellan	116.60	76.75

(b) Execution times by embeddings model.

Model	GPT (s)	Llama (s)
ALBERT	150.72	82.11
BERT	151.94	166.94
DistilBERT	104.65	107.82
BART	432.64	166.94
GPT	181.51	101.80
RoBERTa	174.82	177.49
MiniLM	68.77	65.67

(c) Magellan datasets characteristics and execution times for GPT-3.5 as an auxiliary information model and GPT as an embeddings model.

Dataset	#Rows	#Atts	GPT (s)	Llama (s)
amazon_google_exp	3,226	4	13.26	15.94
beeradvo_ratebeer	3,000	5	21.08	23.60
dblp_acm	2,349	5	25.95	29.28
walmart_amazon	22,076	6	41.60	35.87
fodors_zagats	331	7	56.50	52.37
dblp_scholar	66,774	5	37.54	29.73
itunes_amazon	57,742	9	149.95	124.15

Table 7a shows that execution times depend on the dataset source and its characteristics. Wikidata, with 13–20 attributes, exhibits the longest execution times

(704.90 s for GPT-3.5 and 930.10 s for Llama-3.2). In contrast, smaller datasets with fewer attributes, such as Magellan (4–9 attributes), are processed significantly faster (116.60 s for GPT-3.5 and 76.75 s for Llama-3.2). Table 7b highlights variations in execution times across embedding models. Llama-3.2 demonstrates a range of execution times, from 65.67 s (MiniLM) to 177.49 s (RoBERTa). Similarly, GPT-3.5 processes lightweight models like MiniLM efficiently (68.77 s) but requires substantially more time for models with larger architectures and deeper layers, such as BART (432.64 s). Table 7c shows that execution times for Magellan datasets scale with the number of attributes and rows. Smaller datasets, such as *amazon_google_exp* (13.26 s for GPT-3.5), are processed faster, whereas larger datasets, like *itunes_amazon* (149.95 s for GPT-3.5), take significantly longer due to their higher row counts (57,742 rows) and attribute size (9 attributes).

Discussion. Execution times are influenced by dataset characteristics, generation models, and embedding models. While Llama-3.2 is generally faster, GPT-3.5 performs better on datasets with complex relationships between attributes, such as Wikidata. The number of attributes is the primary factor affecting execution time, as auxiliary information and embeddings are generated for each attribute-instance pair. Lightweight embedding models like MiniLM are efficient but may struggle with datasets requiring more detailed or nuanced analysis. GPT-3.5, although slightly slower, delivers superior performance on tasks involving higher attribute counts or complex data, making it a dependable choice for tasks requiring both accuracy and stability.

4.6 RQ4. Noise Sensibility

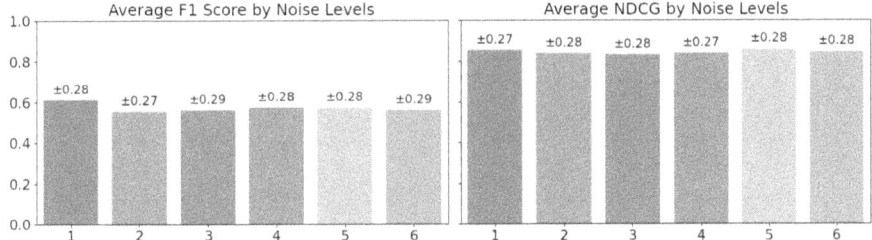

Fig. 3. Impact of noise levels on F1 Score and NDCG across datasets for GPT embeddings model using GPT-3.5 for generation.

To assess robustness, varying levels of noise were introduced across the datasets at the attribute level. These transformations simulate real-world inconsistencies in attribute naming, including typographical changes, structural modifications, and abbreviations. The noise levels range from minimal changes, such as slight typographical errors, to extreme distortions, where attribute names become highly unclear or ambiguous. For example, the attribute *assay_subcellular_fraction* progressively becomes *ASS_SUBCFR* (abbreviation), then *ss_sbclllr_frctn* (aggressive vowel removal), and finally *b_frmt* (complete semantic drift). These perturbations challenge the matching process by reducing or removing semantic cues, reflecting realistic scenarios in schema integration tasks. This approach ensures a rigorous evaluation of the algorithm's ability to adapt to inconsistencies in heterogeneous datasets.

Figure 3 shows the F1 Score and NDCG performance for various noise levels. The evaluation uses GPT-3.5 for embeddings and auxiliary information generation, identified as the most effective combination in RQ1. The F1 Score remains stable, ranging from 0.57 ± 0.29 at level 1 (minimal changes, e.g., typographical errors) to 0.60 ± 0.28 at level 6 (significant distortions). This indicates minimal impact of noise on performance. Similarly, NDCG values show strong consistency, ranging from 0.79 ± 0.28 to 0.80 ± 0.28, with marginal standard deviation differences, demonstrating the solution's ability to maintain ranking quality despite noise.

Discussion. The solution's robustness across different noise levels highlights its effectiveness in handling schema-only and instance-only data sources. The minimal impact of schema-level noise, such as typographical or structural changes, is attributed to GPT's data enrichment process, which contextualizes noise and enhances comprehension. This robustness ensures high performance in both classification and ranking tasks, making the solution well-suited for real-world noisy datasets.

4.7 RQ5. Ablation

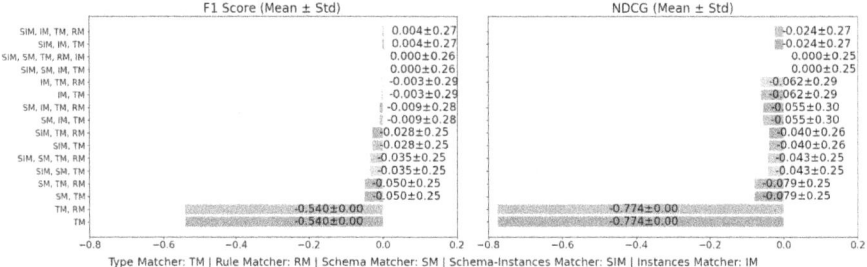

Fig. 4. Impact of matcher ablation on F1 Score and NDCG for the GPT embeddings model across diverse datasets using GPT-3.5 for generation auxiliary information.

The RQ4 experiments analyzed the impact of ablating individual matchers on the framework's performance (F1 Score and NDCG), using GPT-3.5 for embeddings and auxiliary information generation. The findings highlight the critical contribution of specific matchers to the framework's overall performance. Type Matcher was always included as a default component since it serves as a prerequisite for activating other matchers when the type condition is satisfied. Regression analysis identified Instances Matcher and Schema-Instances matcher (Considering Type Matcher) as the most significant contributors, with statistically significant p-values of 0.0282 and 0.0054, respectively (Table 8). These matchers substantially reduced the model's AIC from 497.18 to 279.78, demonstrating improved model quality. The model explained 79% of the variance (McFadden's $R^2 = 0.791$), further confirming the importance of these matchers. Performance analysis of matcher combinations in Fig. 4 revealed that the optimal configuration includes Instances Matcher, Schema-Instances matcher and Type Matcher, which consistently achieved the highest results. This combination produced an F1 Score increase of **0.004 ± 0.27** but an NDCG decrease of **−0.024 ± 0.27**, in comparison with "All matchers" configuration, which scored **0.54 ± 0.26** for F1 and **0.79 ± 0.25** for NDCG.

Discussion. The findings underscore the critical role of Instances Matcher and Schema-Instances matcher in achieving optimal performance. These matchers demonstrated significant contributions not only statistically but also practically, enhancing precision but decreases ranking quality. The observed reduction in AIC suggests improved model parsimony and efficiency, while the slight improvements in F1 Score (+0.004) emphasize the impact of small optimizations on overall performance.

The results also highlight the potential to further improve performance by excluding less impactful matchers, such as Rule Matcher and Schema Matcher, which may detract from the overall quality. This suggests that specific matcher combinations may perform better for certain dataset characteristics, presenting an opportunity to develop recommendations tailored to dataset attributes. These findings align with the hypothesis that matchers interact synergistically, emphasizing the need to adapt matcher combinations dynamically for different data scenarios.

Table 8. Logistic regression results and model statistics for the optimal matcher combination tested on the GPT embeddings model using GPT-3.5 for generation.

Regression Results			Model Stats	
Metric	Value	p-value	Metric	Value
Intercept	-1.96	< 0.001 *	Null Deviance	116.18
instances Matcher	1.82	0.028 *	Residual Deviance	24.27
Schema-Instances matcher	2.63	0.005	AIC	279.78
			McFadden's R^2	0.791

5 Conclusion

We proposed a novel framework for aligning schema-only and instance-only structured data sources, leveraging Schema-Instance matchers and LLMs to generate auxiliary information and enable alignment. Experiments on the Valentine Benchmark showed GPT-3.5 as the best performer for auxiliary generation and embeddings, achieving optimal F1 and NDCG scores. Ablation studies confirmed the importance of Instances Matcher, Schema-Instances Matcher, and Type Matcher in enhancing effectiveness and ranking quality. Our efficiency analysis highlighted the influence of dataset size and complexity on execution time, offering insights for scaling data matching tasks. Future work will explore additional enrichments beyond LLMs, the selection of optimal matcher combinations, and machine learning techniques to improve both performance and efficiency.

References

1. Rahm, E., Bernstein, P.A.: A survey of approaches to automatic schema matching. VLDB J. **10**(4), 334–350 (2001). https://doi.org/10.1007/s007780100057
2. Kellou-Menouer, K., et al.: A survey on semantic schema discovery. VLDB J. **31**, 675–710 (2022). https://doi.org/10.1007/s00778-021-00717-x
3. Madhavan, J., Bernstein, P., Rahm, E.: Generic schema matching with Cupid. In: Proceedings of the VLDB, pp. 49–58 (2001)
4. Zhang, M., et al.: Automatic discovery of attributes in relational databases. In: Proceedings of the SIGMOD, pp. 109–120 (2011). https://doi.org/10.1145/1989323.1989336
5. Aumueller, D., et al.: Schema and ontology matching with COMA++. In: Proceedings of the SIGMOD, pp. 906–908 (2005). https://doi.org/10.1145/1066157.1066283
6. Kired, N.E., et al.: Embedding-based data matching for disparate data sources. In: Wrembel, R., Chiusano, S., Kotsis, G., Tjoa, A.M., Khalil, I. (eds.) DaWaK 2024. LNCS, vol. 14912, pp. 66–71. Springer, Cham (2024). https://doi.org/10.1007/978-3-031-68323-7_5
7. Kellou-Menouer, K., et al.: A Survey on Semantic Schema Discovery. Springer, Cham (2023)
8. Attouche, L., et al.: Optimistic data generation for JSON schema. Trans. Large-Scale Data- Knowl.-Cent. Syst. LVI 119–152 (2024). https://doi.org/10.1007/978-3-662-69603-3_5
9. Adelfio, M.D., Samet, H.: Schema extraction for tabular data on the web. PVLDB **6**(6), 421–432 (2013). https://doi.org/10.14778/2536336.2536343
10. Cui, L., et al.: Tabular data augmentation for machine learning. arXiv:2407.21523 (2024). https://doi.org/10.48550/arXiv.2407.21523
11. Ye, J., et al.: A comprehensive capability analysis of GPT-3 and GPT-3.5 series models. arXiv:2401.03426 (2024). https://doi.org/10.48550/arXiv.2401.03426
12. Grattafiori, A., et al.: The LLaMA 3 herd of models. arXiv:2407.21783 (2024). https://doi.org/10.48550/arXiv.2407.21783
13. Ahmadi, N., et al.: Unsupervised matching of data and text. In: Proceedings of the ICDE, pp. 1058–1070 (2022). https://doi.org/10.1109/ICDE53745.2022.00084

14. Li, H., et al.: On leveraging large language models for enhancing entity resolution. arXiv:2401.03426 (2024). https://doi.org/10.48550/arXiv.2401.03426
15. Suchanek, F.M., Abiteboul, S., Senellart, P.: PARIS: probabilistic alignment of relations, instances, and schema. arXiv:1111.7164 (2011)
16. Zhang, J., et al.: LSM: schema matching using pre-trained language models. In: Proceedings of the ICDE, pp. 2002–2015 (2023)
17. Koutras, C., et al.: Valentine: evaluating matching techniques for dataset discovery. arXiv:2010.07386 (2021). https://doi.org/10.48550/arXiv.2010.07386
18. Demsar, J.: Statistical comparisons of classifiers over multiple data sets. J. Mach. Learn. Res. **7**, 1–30 (2006)

RCIS Forum

Can Llama 3 Accurately Assess Readability? A Comparative Study Using Lead Sections from Wikipedia

José Frederico Rodrigues[1]($\boxtimes$), Henrique Lopes Cardoso[2], and Carla Teixeira Lopes[1]

[1] INESC TEC, Faculdade de Engenharia, Universidade do Porto, Porto, Portugal
{up201807626,ctl}@fe.up.pt
[2] LIACC, Faculdade de Engenharia, Universidade do Porto, Porto, Portugal
hlc@fe.up.pt

Abstract. Text readability is vital for effective communication and learning, especially for those with lower information literacy. This research aims to assess Llama 3's ability to grade readability and compare its alignment with established metrics. For that purpose, we create a new dataset of article lead sections from English and Simple English Wikipedia, covering nine categories. The model is prompted to rate the readability of the texts on a grade-level scale, and an in-depth analysis of the results is conducted. While Llama 3 correlates strongly with most metrics, it may underestimate text grade levels.

Keywords: Readability Assessment · Large Language Models · Llama

1 Introduction

Text clarity is crucial for effective communication, understanding, and learning, particularly for those with lower information literacy. Readability affects how well readers engage with written content, whether in academic, medical, or everyday contexts. Gunning [9] stresses the importance of evaluating readability to ensure students are provided with materials at an appropriate difficulty level, while Manning [12] highlights writing strategies in healthcare to create clear, accessible messages. Complex terminology in fields like law and engineering poses similar challenges. Moreover, readability is core to user experience, especially as generative models are increasingly integrated into systems. Accurately assessing it is important to ensure systems are accessible to readers of varying abilities [16].

Conventional readability metrics generate scores based on elements such as sentence length or word syllables but overlook factors such as content relevance or semantics, as shown in Table 1, which are crucial for a more comprehensive understanding. Despite these limitations, they remain a simple way to estimate text readability. Large language models, however, are emerging as powerful tools

Table 1. Traditional readability metrics

Metric	Features considered
FK [10]	Words per sentence, syllables per word
GF [4]	Words per sentence, complex words (≥ 3 syllables)
SMOG [11]	Number of polysyllables per sentence
ARI [17]	Characters per word, words per sentence
DC [7]	Percentage of difficult words based on a list. Words per sentence
CL [5]	Characters per word, sentences per 100 words
LW [6]	Easy (≤ 2 syllables) and difficult (≥ 3 syllables) words per sentence

FK = Flesch-Kincaid Grade Level; GF = Gunning Fog Index; ARI = Automated Readability Index; DC = Dale-Chall; CL = Coleman-Liau; LW = Linsear Write

in natural language processing, with the potential to accurately assess readability and overcome the conventional metrics' shortcomings.

Llama 3[1], announced on April 18, 2024, is a free model that can be run locally, making it ideal for this investigation due to the high volume of requests involved. We explore Llama 3's performance across multiple domains, comparing it to existing readability metrics. A new dataset is created using lead sections from English Wikipedia (EW) and Simple English Wikipedia (SEW), covering nine categories. We prompt the model to rate the readability on a grade-level scale, and we analyze its correlation with Table 1's readability metrics, which estimate the years of education required to understand a text.

2 Related Work

In studies by Naous et al. [13], Blaneck et al. [3], and Golan et al. [8] LLMs are directly applied for readability assessment. Naous et al. employed both supervised and unsupervised approaches with BERT, mBERT, and XLM-RoBERTa for English and multilingual readability tasks, fine-tuning them on the README++ dataset [13] annotated using Common European Framework of Reference for Languages (CEFR) standards. In other languages, language-specific models like AraBERT(Arabic) and RuBERT(Russian) were applied, and few-shot prompting was explored with GPT-4 and Llama 2. Blaneck et al. investigated German language readability using GBERT and GPT-2-Wechsel in ensemble approaches to enhance performance, while Golan et al. tested ChatGPT's ability to apply traditional readability formulas without relying on annotated datasets. Performance evaluation methods rely on metrics such as Pearson Correlation and Root Mean Squared Error (RMSE), which were used to assess the accuracy of LLM predictions against human-annotated readability levels.

---
[1] https://llama.meta.com/llama3/.

Our research prioritizes investigating the model's behavior across multiple categories using a large dataset. Despite not having readability annotations readily available, we are also not restricted to a smaller scale, such as the CEFR, for rating texts; we can allow the LLM more freedom to rate the lead sections as it sees fit. Notwithstanding, we acknowledge that not additionally evaluating the capacity of the model against human-annotated readability corpora is a limitation. Furthermore, unlike previous studies, we leverage the availability of simpler counterparts for each Wikipedia lead section across every category, allowing us to better understand whether using Llama 3 can be a superior choice over traditional metrics for this task.

3 Dataset Creation and Experimental Setup

Our dataset [14], which is also suitable for the evaluation of text simplification tasks [15], includes lead section pairs from both EW and SEW, covering nine categories. Below, we present an example: the first excerpt is a lead section from English Wikipedia, while the second excerpt is its simplified counterpart from Simple English Wikipedia.

```
Tuition payments, usually known as tuition in American English and as tuition fees in
Commonwealth English, are fees charged by education institutions for instruction or
other services. Besides public spending (by governments and other public bodies),
private spending via tuition payments are the largest revenue sources for education
institutions in some countries. In most developed countries, especially countries in
Scandinavia and Continental Europe, there are no or only nominal tuition fees for
all forms of education, including university and other higher education.
```

```
Tuition payments, usually known as tuition in American English and as tuition fees in
Commonwealth English, are fees charged for students looking for a higher education.
Tuition payments are charged by colleges and universities include costs for lab
equipment, computer systems, libraries, facility upkeep and to provide a comfortable
student learning experience.
```

Despite the existence of several datasets suitable for readability assessment tasks, such as the README++ dataset [13], Newsela [19], and the PLABA dataset [1], none span multiple domains while maintaining consistency for their size and text sources. So, creating a new dataset was deemed necessary to ensure consistency, drawing all texts from the same source across different domains. EW and SEW were selected as the source of the texts for multiple reasons: EW articles are typically written for a general audience but tend to contain complex language, while SEW specifically aims to be more accessible, resulting in a wider range in readability levels across both encyclopedias. Wikipedia covers many topics, allowing the dataset to include many articles from various categories. Lastly, both EW and SEW are freely accessible, making it easy to source many lead sections without licensing issues.

To decide which categories the text samples would be extracted from, we leveraged SEW's category tree and determined the number of article pages of each sub-category directly under the "Everyday Life" and "Knowledge" categories. After analyzing the number of articles per category, we included categories with more than 100,000 articles. These categories were: "Culture", "Education", "Employment", "Entertainment", "Health", "Leisure", "Objects", "Science" and "Time". We traverse a given category and its subcategories to collect page titles. The page title acts as the article's unique identifier across EW and SEW. 10,000 lead section pairs were collected for each of the 9 selected categories, and there are no duplicate titles for each category. Overall, the dataset contains 133,240 unique lead sections and is publicly available in a research data repository[2].

The 8B parameter, instruction fine-tuned Llama 3 model, was chosen because of its smaller size and ability to run on consumer hardware. This model has a context window of 8,192 tokens, which translates to 32,768 characters that can be processed in a single prompt if we assume 4 characters per token. In our dataset, the longest lead section contains 14,945 characters, but most are not over 1,000 characters long, never exceeding Llama's context window capacity. Inference is run on a local NVIDIA RTX 3090 GPU, using the Transformers library [18]. The following system prompt was defined to provide the model a general guideline of its task: "Your role is to rate the readability of texts that are provided to you.". To facilitate processing its responses, the model's temperature was set to a low value, 0.01, to make its replies follow the same format as much as possible. Lowering the temperature minimizes response variability, but due to the nature of readability assessment, this presented itself as an adequate alternative to requesting a specific format through prompt engineering, which proved ineffective, as the model's grading often mismatched its justification for its rating, compromising the validity of the results. The text to assess was provided with the prompt: "Consider the following text: {text}" followed by two new lines and the instruction: "Based on your own assessment, rate its readability on a grade-level scale".

4 Results

The readability assessment task was framed as a classification problem where the readability of texts was categorized into discrete grade levels. With this approach, we can directly map the readability of a text to an educational grade level, as the model was prompted to evaluate the readability of the lead sections by rating them on this scale. In this section, we present the findings of this investigation, organized into four subsections. We pre-processed the model's responses and the scores given by 7 traditional readability metrics, shown in Table 1 and calculated using textstat[3], so as to establish correlations and facilitate comparisons between grade levels. The traditional readability metric scores are floored, meaning a score of 6.7, for example, will correspond to grade level 6.

---

[2] https://rdm.inesctec.pt/dataset/cs-2024-008.
[3] https://textstat.org/.

4.1 Grade Level Distributions

In its response, for both EW and SEW, the model attributes either a grade level to a text or a range spanning up to 3 levels, such as "10th-grade to 12th-grade". Llama provided readability ratings predominantly as ranges rather than single values. Specifically, 80.9% of Llama's responses were in the form of grade level intervals, while the remaining 19.1% were single values. Ranges of values output by Llama never spanned more than three grade levels, and the model's lowest and highest ratings attributed to a lead section were 2nd to 3rd grade and 12th to 14th, respectively. Overall, 53% of all ratings output by Llama were in the interval format of 9th to 10th grade. Higher readability ratings, indicating more complex text, were more common in the Science and Education categories. In contrast, the Leisure category rarely received higher ratings, suggesting that the lead sections in this category are deemed to be written at a relatively lower grade level. On the other hand, lower readability ratings were distributed across all categories more evenly, pointing toward a balanced presence of simpler texts.

While Llama tends to cluster its ratings within narrower intervals, traditional metrics seem to capture more variations in text complexity. The distribution of these scores post-processing is displayed for each metric in Fig. 1. DC's scores, however, escape the trend by tightly clustering around grades 9 and 10, similar to the model.

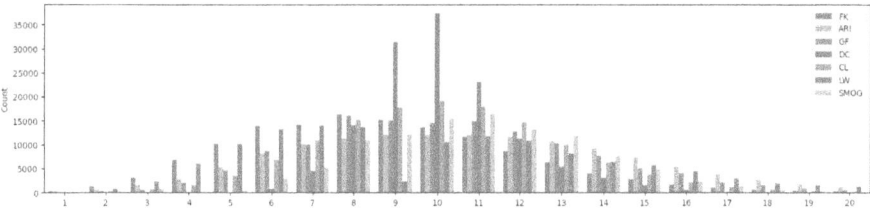

Fig. 1. Overall distribution of readability metric values (1 to 20).

4.2 Deviation Analysis

Llama's ratings in the form of ranges required conversion to single values for meaningful comparison with the processed traditional readability scores. For each LLM rating provided as an interval, we selected the value within the interval that was closest to the grade level given by the metric we were comparing to.

The deviation between a metric's score and the model's rating refers to the difference between the rating provided by the Llama3 model and the readability score given by the traditional metric. To gain insights into how the model's assessment criteria align with established readability measures, we analyzed these deviations. In general, deviations between -3 and 3, displayed in Table 2, account for over 83% of all ratings. Dale-Chall stands out with 93.6% of its scores within this interval. The Dale-Chall metric shows the most significant alignment with the model, followed by Coleman-Liau. Overall, except for the Flesch-Kincaid, with 47% of positive deviations, results indicate that deviations are mostly negative, suggesting that the model is rating lead sections as simpler than the readability metrics convey.

Table 2. Percentage of deviations less than 0, greater than 0, and between -3 and 3.

LLM-Metric	FK	GF	SMOG	ARI	DC	CL	LW
= 0	27.1	27.3	29.4	21.0	**44.0**	32.8	15.3
<0	25.9	**46.6**	**55.9**	**54.1**	37.2	**42.1**	**44.6**
>0	**47.0**	26.1	14.7	24.8	18.8	25.0	40.0
Between -3 and 3	83.2	83.3	88.7	73.2	93.6	88.7	70.5

Bold highlights the highest deviation value/tendency (= 0; < 0; > 0) for each metric

Due to the imbalanced nature of the readability ratings, we report the macro-averaged Mean Absolute Error [2] between the traditional readability metric scores and the model's ratings across all categories, as displayed in Fig. 2. This metric averages the MAE computed for each rating, giving them an equal weight. Notably, The Dale-Chall metric consistently shows one of the lowest errors across most categories, which aligns with the earlier observation of its narrower range and higher alignment with the model's ratings. The SMOG metric also shows relatively low errors in most categories, all of them practically identical, except for the Science and Education categories, where the error is slightly higher at 5.3 and 4.7, respectively. The Objects category seems to be where the error is lowest across most metrics, whereas the category where the error is highest varies.

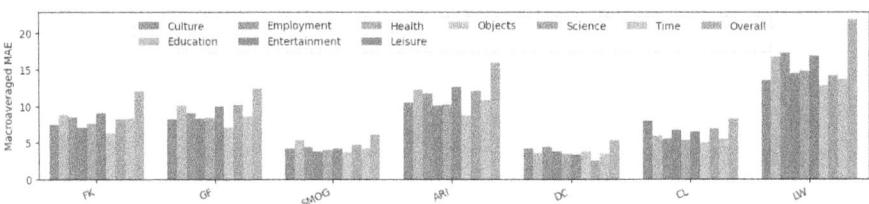

Fig. 2. Macro-averaged Mean Absolute Error between readability metrics and LLM ratings for all categories.

4.3 Correlation Analysis

To complement the analysis, Spearman's Rank Correlation, displayed in the left part of Table 3, is also reported. Overall, the model's predicted readability ratings have a strong positive correlation with all metrics except for DC, which shows a moderate positive association, suggesting a high degree of agreement between Llama and traditional metrics. Out of every metric, Llama's ratings have the strongest positive association with FK, peaking in the Education category. In contrast, the model's correlation is weakest with the DC metric in every category, exhibiting a moderate positive association. DC, however, achieved the lowest macro-averaged MAE among all the metrics. Furthermore, DC is the metric with the highest number of ties with the model's ratings, resulting in a much larger number of tied rank situations when calculating Spearman's rank correlation coefficient, which could impact the measure's accuracy.

Table 3. Left: Spearman's Rank Correlation between LLM ratings and readability metrics for all categories. **Right:** Percentage of cases where readability ratings for SEW were equal or lower than ratings for EW.

Category	Spearman's Correlation							Percentage of cases						
	FK	GF	SM	AR	DC	CL	LW	LLM	FK	GF	SM	AR	DC	CL	LW
Culture	0.81	0.76	0.76	0.77	0.51	0.71	0.78	96.8	87.7	87.0	93.0	86.9	74.3	83.7	86.1
Education	**0.82**	0.78	0.77	0.79	0.60	0.74	0.78	96.6	91.4	89.8	94.2	90.9	85.3	88.7	88.9
Employment	0.79	0.79	0.77	0.79	0.47	0.76	0.79	94.1	91.6	90.7	93.3	91.2	80.9	88.2	**90.4**
Entertainment	0.76	0.72	0.74	0.73	0.40	0.66	0.75	95.1	81.1	79.4	91.0	78.4	59.0	72.7	79.9
Health	0.79	0.78	0.76	0.78	0.61	0.75	0.77	96.0	**92.3**	**90.9**	94.0	**91.9**	84.7	**90.1**	89.9
Leisure	0.73	0.71	0.68	0.72	0.37	0.65	0.72	95.3	87.2	85.8	92.1	86.4	73.7	82.0	85.3
Objects	0.73	0.76	0.78	0.76	0.55	0.69	0.73	94.7	82.6	83.4	93.4	84.0	75.1	86.3	79.7
Science	0.81	**0.80**	**0.81**	**0.82**	**0.65**	**0.77**	0.78	**97.7**	92.1	90.8	**94.8**	90.9	**87.3**	89.5	89.2
Time	0.80	**0.80**	0.79	0.81	0.48	0.74	**0.80**	96.2	87.9	88.7	94.4	75.0	87.4	86.8
Overall	0.79	0.78	0.78	0.78	0.51	0.74	0.76	95.7	88.5	87.8	93.4	88.2	78.3	86.2	86.5

Bold highlights the highest value per column.
SM= SMOG; AR = ARI.

Very strong positive correlations are observed across most metrics for the Science and Time categories, and for each metric except Linsear-Write, the strongest positive correlation is observed in the Science category. In contrast, Leisure is the category where each metric displays its weakest correlation, closely followed by the Entertainment category.

4.4 English Wikipedia Vs Simple English Wikipedia

We compare readability assessments between standard and simplified Wikipedia lead sections since these simplified versions are human-generated and could include aspects that aren't considered by traditional metrics. We determine the percentage of cases where their ratings for SEW sections were equal to or lower than their ratings for EW. Results are shown in the right part of Table 3. Llama's ratings show the highest percentages across most categories, suggesting it rarely deems SEW lead sections as more complex. These results begin to showcase how an LLM could be a better choice over traditional metrics to assess readability.

Metrics such as FK or SMOG rely on surface-level features like word length, sentence length, or syllable count, while a large language model can leverage a contextual understanding, which should align more closely with human judgments. Employing an LLM incurs a greater cost than computing these metrics, but it could be justified in situations where determining if a text's content is easier to understand is given more importance than determining if it is easier to read. Most metrics also display high percentages across all domains, validating the model's assessment. There is, however, a considerable difference in the DC percentages between the Entertainment and Science categories.

5 Conclusions

To study Llama 3's performance in the task of readability assessment, we create a new dataset spanning multiple categories. We explore the distribution and characteristics of the model's output and compare it with scores from traditional readability metrics. Llama tends to grade texts with a level interval instead of a single grade level. It correlates most strongly with FK and weakest with DC. It tends to grade texts as more readable than traditional metrics, except for FK. Lastly, Llama rarely determines SEW sections as more complex than their EW counterpart, surpassing traditional readability metrics when it comes to distinguishing simple from complex texts, indicating that it does not only rely on surface-level features. Overall, results suggest that Llama 3 can accurately assess readability while overcoming weaknesses inherent to traditional metrics. This can be particularly relevant for educators, policymakers involved in curriculum development, and especially developers of educational tools, as a more nuanced alternative to standard methods that could help pave the way for stronger learning tools and platforms.

Acknowledgments. This research is funded by the Department of Informatics Engineering and the Master's program in Informatics and Computing Engineering at the Faculty of Engineering, University of Porto. National Funds also support it through the Portuguese funding agency, FCT - Fundação para a Ciência e a Tecnologia, within project LA/P/0063/2020 (DOI: 10.54499/LA/P/0063/2020). Additional support comes from Base (UIDB/00027/2020) and Programmatic (UIDP/00027/2020) Funding of the Artificial Intelligence and Computer Science Laboratory (LIACC), funded by national funds through FCT/MCTES(PIDDAC). Furthermore, we acknowledge the Institute for Systems and Computer Engineering, Technology and Science (INESC TEC) for granting access to computational resources, including the machine used for running the LLM experiments.

References

1. Attal, K., Ondov, B., Demner-Fushman, D.: A dataset for plain language adaptation of biomedical abstracts. Sci. Data **10**, 8 (2023). https://doi.org/10.1038/s41597-022-01920-3

2. Baccianella, S., Esuli, A., Sebastiani, F.: Evaluation measures for ordinal regression, pp. 283–287 (2009). https://doi.org/10.1109/ISDA.2009.230
3. Blaneck, P.G., Bornheim, T., Grieger, N., Bialonski, S.: Automatic readability assessment of German sentences with transformer ensembles. arXiv preprint arXiv:2209.04299 (2022)
4. Brucker, C.: The gunning's fog index (or fog) readability formula. https://readabilityformulas.com/the-gunnings-fog-index-or-fog-readability-formula/. Accessed 31 May 2024
5. Coleman, M., Liau, L.: A computer readability formula designed for machine scoring (1975)
6. CSUN: Readability helps the level (2006). http://www.csun.edu/~vcecn006/read1.html. Accessed 31 May 2024
7. Dale, E., Chall, J.S.: A formula for predicting readability: instructions. Educ. Res. Bull. **27**(2), 37–54 (1948). http://www.jstor.org/stable/1473669
8. Golan, R., et al.: ChatGPT's ability to assess quality and readability of online medical information: evidence from a cross-sectional study. Cureus (2023). https://doi.org/10.7759/cureus.42214
9. Gunning, T.G.: The role of readability in today's classrooms. Top. Lang. Disord. **23**, 175–189 (2003). https://doi.org/10.1097/00011363-200307000-00005
10. Kincaid, J.P., Fishburne Jr., R.P., Rogers, R.L., Chissom, B.S.: Derivation of new readability formulas (automated readability index, fog count and flesch reading ease formula) for navy enlisted personnel. http://library.ucf.edu
11. Laughlin, G.: Smog grading-a new readability formula. J. Read. **12**(8), 639–646 (1969)
12. Manning, D.T.: Writing readable health messages. Public Health Rep. **96**(5), 464–465 (1981). https://api.semanticscholar.org/CorpusID:39039337
13. Naous, T., Ryan, M.J., Lavrouk, A., Chandra, M., Xu, W.: ReadMe++: benchmarking multilingual language models for multi-domain readability assessment (2023)
14. Rodrigues, J.F., Teixeira Lopes, C., Lopes Cardoso, H.: Wikipedia and simple Wikipedia lead section pairs for nine categories (2024). https://doi.org/10.25747/4VC9-ZS43. [Data set]. INESC TEC
15. Rodrigues, J.F., Teixeira Lopes, C., Lopes Cardoso, H.: Evaluating llama 3 for text simplification: a study on Wikipedia lead sections. In: Companion Proceedings of the ACM Web Conference 2024, WWW 2025. Association for Computing Machinery (2025). https://doi.org/10.1145/3701716.3715467
16. Roegiest, A., Pinkosova, Z.: Generative information systems are great if you can read. In: Proceedings of the 2024 Conference on Human Information Interaction and Retrieval, CHIIR 2024, pp. 165–177. Association for Computing Machinery, New York (2024). https://doi.org/10.1145/3627508.3638345,
17. Smith, E.A., Senter, R.J.: Automated readability index. AMRL-TR. Aerospace Medical Research Laboratories (U.S.), pp. 1–14 (1967)
18. Wolf, T., et al.: Transformers: state-of-the-art natural language processing. In: Liu, Q., Schlangen, D. (eds.) Proceedings of the 2020 Conference on Empirical Methods in Natural Language Processing: System Demonstrations, pp. 38–45. Association for Computational Linguistics, Online (2020). https://doi.org/10.18653/v1/2020.emnlp-demos.6, https://aclanthology.org/2020.emnlp-demos.6
19. Xu, W., Callison-Burch, C., Napoles, C.: Problems in current text simplification research: new data can help. Trans. Assoc. Comput. Linguist. **3**, 283–297 (2015). https://doi.org/10.1162/tacl_a_00139

Research Challenges in Routine Optimization for Synthesizing Software Robots

J. L. Alonso-Rocha(✉), A. Martínez-Rojas, A. Jiménez-Ramírez, and J. G. Enríquez

Department of Computer Languages and Systems, University of Seville, Avenida Reina Mercedes, s/n. 41012, Seville, Spain
{jalonso2,amrojas,ajramirez,jgenriquez}@us.es

Abstract. Robotic Process Automation (RPA) leverages software robots to streamline repetitive rules-based tasks, enhancing efficiency and reducing errors. Advances in Robotic Process Mining (RPM) and Task Mining (TM) enable the identification and segmentation of automatable routines from user interaction logs. However, despite these advances, significant gaps and challenges persist in various stages of the RPM pipeline. These challenges hinder the effective discovery and optimization of routines, limiting the efficiency and robustness of the resulting software robots. This work systematically identifies and organizes these challenges in a structured framework. Drawing on previous research, we define four key categories of routine optimization issues. This classification provides a foundation for analyzing and addressing existing gaps, offering a broad perspective of the complexities involved. By developing and applying these categories, we provide a flexible framework to guide further research on routine optimization for improving software robots.

Keywords: Robotic Process Automation · UI Logs · Process Mining · Task Mining

1 Introduction

Robotic Process Automation (RPA) allows software agents to mimic user actions on graphical user interfaces (GUIs) to automate repetitive and trivial business tasks [1]. Over the last decade, RPA has been increasingly complemented by Robotic Process Mining (RPM) and Task Mining (TM) techniques, which help identify and segment automatable routines from UI (User-interface) logs [4,15]. By capturing and analyzing user actions, RPM and TM provide valuable

---

This research was supported by the EQUAVEL project PID2022-137646OB-C31, funded by MICIU/AEI/10.13039/501100011033 and by ERDF, EU.; the grant FPU20/05984 funded by MICIU/AEI/10.13039/501100011033 and by ESF+.

© The Author(s), under exclusive license to Springer Nature Switzerland AG 2025
J. Grabis et al. (Eds.): RCIS 2025, LNBIP 548, pp. 98–108, 2025.
https://doi.org/10.1007/978-3-031-92471-2_7

insights into how these routines unfold in practice, forming the foundation for automated solutions based on software robots that promise higher efficiency and reduce human errors [5,8]. Despite advances in RPM and TM, issues remain in routine optimization. UI logs from real scenarios often contain inconsistent user interactions, such as redundant actions, errors, or deviations from intended processes [13]. These inconsistencies complicate the derivation of clear, automatable routines that accurately represent an optimized process. Addressing these issues allows for obtaining a clean, structured UI log that improves software robot generation. To this end, previous studies have contributed to specific aspects of routine discovery and optimization:

Leno et al. [9] explore the discovery of automatable routines from UI logs focusing on log segmentation, frequent pattern extraction, and the synthesis of executable specifications for RPA scripts. The work highlights challenges such as noise sensitivity, overlapping routines, and the inability to detect conditional behaviors inside the log.

Agostinelli et al. [2] presents a method for synthesizing software robots by analyzing user-executed routines recorded in presegmented UI logs. The approach generates cross-platform RPA scripts and accommodates intermediate manual inputs. However, it relies on clean and controlled logs and highlights segmentation and routine variability as key challenges for real-world applications.

Although the state-of-the-art provides valuable contributions to routine identification, new challenges emerge across the RPM and RPA pipeline. A recent systematic review by El-Gharib et al. [5] highlights the persistent and diverse challenges within Process Mining (PM) and RPA, emphasizing the need to continue researching this topic to address these open issues.

Building on this foundation, our work offers a classification framework that organizes known challenges such as noise in logs, overlapping routines, and conditional interactions-which affect routine optimization. The framework outlines gaps through brief descriptions and cases, enabling their potential resolution. This contributes to cleaner UI logs, improving future software robot generation.

Thus, the rest of this paper is structured as follows: Sect. 2 reviews the state-of-the-art. Section 3 presents a classification framework for categorizing current gaps and identified challenges. Section 4 discusses the study's limitations and proposes directions for future research. Finally, Sect. 5 concludes the paper by summarizing key findings and their implications.

2 Related Work

This section conducts a state-of-the-art review to identify the main challenges in routine optimization. We followed the recommendations and guidelines proposed by Kitchenham et al. [7] to achieve this goal. This method consists of 3 phases, i.e., *planning* where search conditions and criteria are defined; *conducting*, applying these criteria to identify primary studies; and *reporting*, evaluating the insights from them.

Regarding the *planning* phase, Scopus[1] was selected as the primary database due to its widespread adoption and reliability in academic research [3]. Although our search aimed to identify primary studies directly addressing routine optimization, we found no papers specifically focused on this area. To address this limitation, we adjusted our search criteria to adopt a more general approach. This scope allowed us to identify primary studies that present problems that can be addressed through routine optimization. Thus, the query that was executed in the Scopus digital library was: (**"Robotic Process Automation" OR RPA**) **AND ("Process Mining" OR "Robotic Process Mining" OR RPM OR "Task Mining" OR TM) AND ("challenge*" OR "review")**.

Moreover, various inclusion (I) and exclusion (E) criteria (C) were established to filter the results, ensuring thematic relevance and high quality:

- **IC01.** Studies were obtained by configuring the search query to focus on the ones that include the specified terms in their title, abstract, or keywords.
- **IC02.** Selection of studies that explicitly identify future research directions, gaps, or challenges related to routine optimization.
- **EC01.** Exclusion of studies before 2020. Studies from earlier years were excluded to focus on recent advancements and ensure relevance to current research. This selection aligns with an exponential growth in RPA-related topics. Enríquez et al. [6] illustrate this trend through a graph showing a sharp increase in studies through 2019.
- **EC02.** Exclusion of studies not published in journals or conferences.

The *conducting* phase involved the execution of the search indicated in the planning phase and filtering based on the inclusion and exclusion criteria. The query in Scopus as of March 2025 (**IC01**) retrieved a total of 38 studies containing the specified search terms. Applying **EC01** excluded 4 studies before 2020, reducing the count to 34. Next, **EC02** removed 10 studies, leaving 24. As the last step, **IC02** narrowed the selection to 2 primary studies related to routine optimization. These steps are illustrated in Fig. 1.

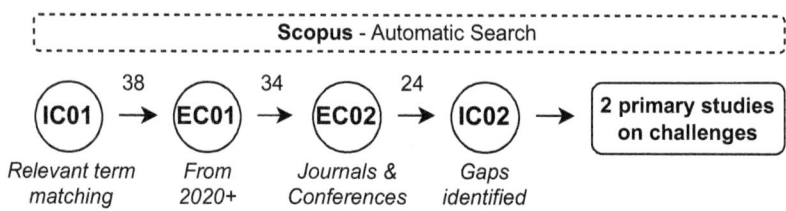

Fig. 1. Visual summary of sequential search criteria application in Scopus.

Finally, in the *reporting* phase, we analyzed the results obtained after the conducting phase. Thus, the two primary studies identified form the basis of our

---

[1] https://www.scopus.com/sources.

research. Each addresses aspects of routine optimization in the context of RPA and RPM. Below, we discuss their contributions and relevance in detail.

Leno et al. [10] propose an RPM pipeline comprising three phases: UI log preprocessing, candidate routine identification, and executable routine discovery. Their methodology includes noise filtering, segmentation, and simplification to extract automation routines and generate platform-specific scripts. The study highlights challenges across the pipeline, focusing on the complexities of achieving full automation. Later work by Leno et al. [9] expands this foundation by identifying new gaps, further enriching the discussion on RPM challenges.

El-Gharib et al. [5] present the most recent systematic literature review (SLR) in our research domain, examining the integration of PM with RPA to enhance implementation. Their study focuses on techniques, tools, and challenges, highlighting issues in data gathering, preprocessing, and the interaction between these disciplines. It introduces a unified framework that addresses challenges arising from their integration for further exploration.

Finally, while these proposals address challenges more generally, our framework structures routine optimization challenges into actionable dimensions, bridging specific gaps and broader perspectives to enable flexible solution exploration.

3 Approach

This section introduces the classification framework and its use to address the problems identified in the literature. Section 3.1 explains its construction and categories, while Sect. 3.2 demonstrates its application to relevant scenarios.

3.1 Methodology

The proposed framework stems from a systematic analysis of prior studies and practical observations. We build on the SLR by El-Gharib et al. [5], which integrates PM and RPA challenges into a shared conceptual model, but it remains limited by real-world UI log complexity and the diverse scenarios of routine optimization. To address more specific issues, we focus on the gaps identified by Leno et al. [9], i.e., the most recent work closely aligned with this field. Although other research (e.g., Syed et al. [14]) also examines routine optimization, Leno's study highlights unresolved concerns well-suited to our framework. Additionally, we incorporated a set of problems from previous works [11–13], described in a Technical Report[2], thereby enriching and updating Leno's findings. By combining these insights, we identified common factors that, together with El-Gharib et al.'s conceptual base [5], enabled us to define a set of dimensions for classifying routine optimization gaps into meaningful categories.

- **D1. Primary Concern:** Identifies the core issue behind each challenge, such as log inconsistencies or ambiguous routine boundaries. By clarifying

---

[2] Technical Report available at: https://doi.org/10.5281/zenodo.14733044.

the main problem area, this dimension ensures that subsequent actions target the fundamental obstacle instead of merely treating superficial symptoms.
- **D2. Required Interventions:** Outlines the transformations needed to address the identified concern, e.g., reorganizing events, reconciling data, or adapting conditional behaviors. Detailing these required operations connects the conceptual challenge with practical solution methods.
- **D3. Target Outcome:** Establishes the intended result of resolving the challenge, such as achieving a coherent workflow or filtering outliers. Emphasizing the final goal ensures that solutions advance broader objectives in routine optimization rather than solving isolated issues.

Next, we present the proposed categories that form the core of our classification framework:

- *DHD - Data Handling and Deriving:* Centers on managing multiple data inputs and identifying recurring patterns in user interactions (D1). It addresses scenarios where data entries, often dispersed or repeated, require systematic merging or processing (D2). Involves deriving data from logs to support routine identification and ensure consistency (D3).
- *RCS - Routine Consistency and Structuring:* Focuses on identifying, separating, and optimizing user interactions within automatable routines (D1). Typical approaches include segmenting logs or excluding irrelevant events that hinder routine clarity (D2). The goal is to create a coherent, error-free sequence reflecting the intended interaction flow (D3).
- *CFA - Component Flow and Adapting:* Deals with dynamic states and conditional interactions in the UI (D1). Transformations may involve detecting changes in UI components (hereafter referred to as "components") status or handling dependencies between events (D2). This enables routines to accommodate various interface conditions, including conditional triggers (D3).
- *FDC - Filtering and Data Cleaning:* Addresses the elimination of irrelevant events and log cleaning (D1). Techniques include filtering, removing outliers, and consolidating fragmented user inputs (D2). This would allow us to obtain a log that is free from irrelevant actions to improve routine accuracy (D3).

3.2 Applying the Classification Framework to Case Studies

This section presents the case studies developed based on the challenges that shaped our research. Each case study is described with its specific gap and an illustrative example. To effectively organize the routine optimization challenges they address, case studies are grouped using the classification framework described in Sect. 3.1, i.e., primary categories for high-level grouping and secondary categories for capturing additional dimensions of the challenges.

- **DHD-1: Aggregating Multi-Field Data**
    - *Secondary Category:* Routine Consistency and Structuring.
    - *Gap:* Challenges in handling multiple data inputs aggregated across different fields during a single interaction.

- *Sample Scenario:* In an address form, a user enters "Street", "City", and "Postal Code" in a single action. The logs link multiple data entries to this interaction, requiring grouping and optimization to treat these fields as a unified unit, merging related events to refine the routine structure.
- **DHD-2: Detecting Recurrent Text Entries**
  - *Secondary Category:* Filtering and Data Cleaning.
  - *Gap:* Identifying frequently recurring text entries in specific fields to optimize routine discovery.
  - *Sample Scenario:* Users repeatedly enter the same email address in a log-in form. Detecting this pattern identifies repetitive behaviors for automation, enabling software robots to set default values by filtering recurring text inputs across routine variants.
- **DHD-3: Selecting Component-Specific Events**
  - *Secondary Category:* Component Flow and Adapting.
  - *Gap:* Selecting the best events for components differentiated by their positions (e.g., coordinates) in a form.
  - *Sample Scenario:* In a single form view, users interact with multiple text input fields like "Name" and "Company," which, despite being of the same type, serve distinct purposes. The task is to group events by component and ensure they are correctly attributed to their respective fields, avoiding ambiguity during automation.
- **DHD-4: Optimizing Click Events Across Variants**
  - *Secondary Category:* Filtering and Data Cleaning.
  - *Gap:* Distinguishing the most relevant click events across routine variants by analyzing interactions with components.
  - *Sample Scenario:* In a process with multiple routine variants, clicks on "Add Item" buttons are reviewed to identify the best component interaction event based on criteria like proximity or relevance. Then, less optimal click events are discarded.

- **RCS-1: Segmenting Overlapping Routines**
  - *Secondary Category:* Filtering and Data Cleaning.
  - *Gap:* Requires separating events in unsegmented logs to accurately identify and assign actions to routines.
  - *Sample Scenario:* In a project management application, a user fills out a "New Task" form while simultaneously navigating the main menu to review another project. Both activity clicks and data input are mixed in the logs, necessitating proper identification and separation.
- **RCS-2: Ensuring Semantic Equivalence in Routines**
  - *Secondary Category:* Data Handling and Deriving.
  - *Gap:* Identifying equivalent actions in routines while excluding ineffective events to ensure routines reflect meaningful interactions.
  - *Sample Scenario:* Users clicking a disabled "Save" button generate non-actionable events. The task is to exclude these events, ensuring routine equivalence considers only functional interactions.
- **RCS-3: Discovering Flexible Interaction Patterns**

- *Secondary Category:* Component Flow and Adapting.
- *Gap:* Difficulties in identifying recurring interaction patterns that occur in varying orders but represent the same routine.
- *Sample Scenario:* In a checkout process, a user may input the "Shipping Address" and then the "Payment Method" or vice versa. Logs record these interactions in different sequences, requiring recognition that both represent the same optimized routine. Interactions with components could help to identify routines effectively.

- **CFA-1: Managing Component States**
    - *Secondary Category:* Routine Consistency and Structuring.
    - *Gap:* Capturing and analyzing state changes (e.g., enabled/disabled) to determine their impact on subsequent interaction steps.
    - *Sample Scenario:* In forms with components like "Submit" buttons, the button remains disabled until the user checks a "Terms and Conditions" box. Tracking these state changes is essential to understand how clicks and interactions depend on the component's state to shape the routine.
- **CFA-2: Enriching Contextual Keyboard Inputs**
    - *Secondary Category:* Filtering and Data Cleaning.
    - *Gap:* Linking text input events with preceding clicks to enrich contextual understanding of the input's intent.
    - *Sample Scenario:* A user clicks on a text field before typing, and the preceding click event is linked to the corresponding text input, providing context about the component being interacted with. Merging this information would streamline the routine by removing redundant events and consolidating interactions.
- **CFA-3: Discovering Conditional Interactions**
    - *Secondary Category:* Data Handling and Deriving.
    - *Gap:* Identifying dependencies between components to analyze how actions or states in one trigger or modify behaviors in others.
    - *Sample Scenario:* In a registration form, selecting "Yes" for "Subscribe to Newsletter" enables an additional field for email input. Identifying such conditional interactions would involve deriving relationships between components and understanding how they enable each other.

- **FDC-1: Filtering Accidental Interactions**
    - *Secondary Category:* Routine Consistency and Structuring.
    - *Gap:* Removing irrelevant or accidental interactions that do not contribute to meaningful actions in the routine.
    - *Sample Scenario:* A user clicks outside the interactive area of a component, producing an irrelevant event. Filtering these accidental interactions would streamline the routine by removing unnecessary actions.
- **FDC-2: Streamlining Keyboard Events**
    - *Secondary Category:* Routine Consistency and Structuring.

- *Gap:* Addressing duplicate or incomplete keyboard events caused by user errors, such as deleting and retyping text.
- *Sample Scenario:* A user types text into a field, deletes it and retypes. Redundant events could be filtered and removed, potentially refining the routine by ensuring that only relevant interactions are retained.

– **FDC-3: Handling Special Keyboard Inputs**
- *Secondary Category:* Data Handling and Deriving.
- *Gap:* Filtering out or correctly interpreting special key combinations (e.g., 'Ctrl+C', 'Ctrl+V') that may interfere with regular text input.
- *Sample Scenario:* During text input, users paste content using 'Ctrl+V'. Capturing and storing such key combinations as separate data can enrich future robot automation.

The resolution of these challenges produces a cleaner UI log by reducing inconsistencies, refining event selection, and improving segmentation. Capturing conditional dependencies and filtering irrelevant interactions further minimizes noise. As a result, the UI log reflects an optimized routine with the most relevant events, enhancing the extracted process and enabling the generation of more accurate and reliable software robots.

4 Discussion

This section presents the strengths of our framework, discusses limitations, and outlines future research directions in routine optimization.

Our research introduces a framework for categorizing challenges in the RPM pipeline, providing a structured and reusable foundation for addressing issues. Unlike previous approaches that address challenges in isolation [5] or limit them to specific pipeline phases [9,10], our framework is more flexible and adaptable. The categorization of challenges clarifies research directions and supports the structured development of future studies. On the one hand, CFA challenges might advance understanding of how component states provide new insights into the flow of events within routines. On the other hand, FDC challenges can improve the detection and handling of outliers in routines.

A potential limitation identified is that the set of challenges may be restricted due to its non-exhaustive scope. However, this does not detract from the validity or utility of the proposed framework. Rather than aiming for a comprehensive classification framework, this initial set of challenges demonstrates the framework's functionality and potential. Future research can build on this foundation by identifying additional issues and refining its structure as new insights emerge. A more specific approach to defining the classification framework could clarify the challenges in this field, potentially uncovering new categories and offering a deeper understanding of routine optimization issues.

Using our observations of the classified challenges, we propose exploring event-level analysis to extend the scope beyond high-level abstraction. Existing methodologies often focus on segmenting logs and identifying automatable routines. However, they often overlook optimizing individual events within those routines. By addressing this gap, researchers can investigate how detailed event-level analysis can contribute to constructing optimized routines. This low-level approach involves analyzing and selecting individual events across routine variants based on specific characteristics derived from the logs.

Using the DHD-4 case study as an example, enriched events from UI logs can be used to calculate distances between user clicks and target components through screen coordinates. This approach enables the selection of the most accurate events by prioritizing those with the shortest distance between the click and the target component. Martinez-Rojas et al. [12] demonstrate the feasibility of this method. They show how UI logs can be enhanced with screen data, providing additional parameters for event-level analysis. This enhancement supports the construction of routines based on optimized events, allowing a future robot to replicate actions with greater precision.

5 Conclusion

This study introduces a systematic classification framework to address key issues in routine optimization within RPM. Despite advances in this discipline, the variability of recorded processes within UI logs continues to introduce new challenges across different areas, such as routine identification and segmentation or filtering irrelevant data. Moreover, existing studies often address these challenges in isolation, framing them narrowly as specific problems or linking them to particular RPM pipeline phases. This fragmented approach complicates their resolution and hinders progress toward developing precise software robots.

To address these issues, we analyzed gaps identified in the literature and combined them with insights from our experimental problems. This led to the definition of three core dimensions, i.e., Primary Concern, Required Interventions, and Target Outcome, which provided a foundation for categorizing routine optimization challenges. To validate this approach, we applied it to case studies aligned with the identified challenges, each linked to specific gaps.

These case studies reveal key obstacles in routine optimization. Noise in logs obscures meaningful interactions. It can manifest in different ways, requiring effective detection and filtering algorithms. Overlapping routines complicate the separation of distinct events from different processes. Addressing this requires methods to identify and isolate specific routines accurately. Dynamic and conditional interactions, such as state changes or dependencies between components, demand precise recognition. Understanding how state-based component interactions influence the flow of events provides valuable insights into how routines are structured. Variability in user behavior-such as differing input sequences, repeated actions, or inconsistent patterns-underscores the need for solutions that can adapt to these inconsistencies.

Finally, we propose two future research lines. First, exploring event-level analysis could optimize individual events within routines, enabling the creation of more precise and efficient software robots. Second, expanding the classification framework with new challenges could refine its structure, potentially uncovering new categories and offering a deeper understanding of routine optimization issues. Together, these directions strengthen the RPM lifecycle, particularly its final stages, leading to improved software robots with more effective automation.

References

1. van der Aalst, W., Bichler, M., Heinzl, A.: Robotic process automation. Bus. Inf. Syst. Eng. **60**(4), 269–272 (2018)
2. Agostinelli, S., Lupia, M., Marrella, A., Mecella, M.: SmartRPA: a tool to reactively synthesize software robots from user interface logs. In: Nurcan, S., Korthaus, A. (eds.) CAiSE 2021. LNBIP, vol. 424, pp. 137–145. Springer, Cham (2021). https://doi.org/10.1007/978-3-030-79108-7_16
3. Baas, J., Schotten, M., Plume, A., Côté, G., Karimi, R.: Scopus as a curated, high-quality bibliometric data source for academic research in quantitative science studies. Quant. Sci. Stud. **1**(1), 377–386 (2020)
4. Dumas, M., La Rosa, M., Leno, V., Polyvyanyy, A., Maggi, F.M.: Robotic process mining. Process Mining Handb. **448**, 468–491 (2022)
5. El-Gharib, N.M., Amyot, D.: Robotic process automation using process mining - a systematic literature review. Data Knowl. Eng. **148** (2023)
6. Enríquez, J.G., Jiménez-Ramírez, A., Domínguez-Mayo, F.J., García-García, J.A.: Robotic process automation: a scientific and industrial systematic mapping study. IEEE Access **8**, 39113–39129 (2020)
7. Kitchenham, B., Brereton, O.P., Budgen, D., Turner, M., Bailey, J., Linkman, S.: Systematic literature reviews in software engineering-a systematic literature review. Inf. Softw. Technol. **51**(1), 7–15 (2009)
8. Leno, V., Augusto, A., Dumas, M., La Rosa, M., Maggi, F.M., Polyvyanyy, A.: Identifying candidate routines for robotic process automation from unsegmented ui logs. In: 2020 2nd International Conference on Process Mining (ICPM), pp. 153–160. IEEE (2020)
9. Leno, V., Augusto, A., Dumas, M., La Rosa, M., Maggi, F.M., Polyvyanyy, A.: Discovering data transfer routines from user interaction logs. Inf. Syst. **107**, 101916 (2022)
10. Leno, V., Polyvyanyy, A., Dumas, M., La Rosa, M., Maggi, F.M.: Robotic process mining: vision and challenges. Bus. Inf. Syst. Eng. **63**, 301–314 (2021)
11. Martínez-Rojas, A., Alonso-Rocha, J., Jiménez-Ramírez, A., Enríquez, J.: From screenshots to process models: improving activity identification through screen text. In: International Conference on Business Process Management, pp. 125–137. Springer, Heidelberg (2024). https://doi.org/10.1007/978-3-031-70445-1_8
12. Martínez-Rojas, A., Jiménez-Ramírez, A., Enríquez, J.G., Reijers, H.A.: A screenshot-based task mining framework for disclosing the drivers behind variable human actions. Inf. Syst. **121**, 102340 (2024)

13. Martínez-Rojas, A., Jiménez-Ramírez, A., Enríquez, J.G., Reijers, H.A.: Analyzing variable human actions for robotic process automation. In: International Conference on Business Process Management, pp. 75–90. Springer, Heidelberg (2022). https://doi.org/10.1007/978-3-031-16103-2_8
14. Syed, R., et al.: Robotic process automation: contemporary themes and challenges. Comput. Ind. **115**, 103162 (2020)
15. Urabe, Y., Yagi, S., Tsuchikawa, K., Oishi, H.: Task clustering method using user interaction logs to plan RPA introduction. In: Polyvyanyy, A., Wynn, M.T., Van Looy, A., Reichert, M. (eds.) BPM 2021. LNCS, vol. 12875, pp. 273–288. Springer, Cham (2021). https://doi.org/10.1007/978-3-030-85469-0_18

Mining for Meaning: Ontology-Aware Process Mining Methods Through Knowledge Patterns

Riley Moher[(✉)] and Michael Gruninger

Department of Mechanical and Industrial Engineering, University of Toronto, Ontario M5S 3G8, Canada
riley.moher@mail.utoronto.ca

Abstract. Process mining has emerged as a critical practice for understanding business processes through data-driven analysis. Practitioners necessarily apply diverse process and domain knowledge to guide their analyses. However, these practices are often ad-hoc and informal, failing to formalize the process knowledge being applied. While various formal methods including temporal logic have been applied to process mining, they fail to acknowledge fundamental ontological commitments of processes. Formal ontology for processes, on the other hand, are difficult to integrate into a process mining pipeline, lacking a data-driven grounding. We thus introduce process meaning patterns as a formal declarative framework to capture process knowledge being applied in process mining, based on first order logic ontology patterns. We demonstrate our framework's ability to semi-automatically infer process knowledge motivated by real applications from Volvo IT Belgium. This paper also accompanies a preliminary implementation of the framework using Python and Datalog.

Keywords: process mining · ontology · logic programming · datalog

1 Introduction

Business processes are a nuanced and significant value engine for enterprises. Process mining [1,2] has enabled a bottom-up and data-driven approach to understanding and analyzing these processes, leveraging event logs to uncover, understand, and optimize complex business processes across diverse domains. Given the accelerating volume of enterprise process data, efficient and precise interpretation of this data has become a critical opportunity. However, project budgets, rushed timelines, and heterogeneity of data introduce significant constraints in practice, leading to a reliance on ad-hoc and informal methodologies [3,4]. These constraints limit the depth of process analysis and results in critical process knowledge being opaque and unsuitable for re-use or diagnosis.

Formal methods including upper ontologies [5–7] promise a more formal and structured approach to handle these issues, employing formal logic to represent and reason about fundamental process concepts. However, while these ontologies are expressive and well-formalized, they lack grounding in real enterprise data and lack clear applications for process mining pipelines. While applications of formal logic [8–10], knowledge graphs [11,12], and declarative models [8,9] have been applied to process mining, they focus on narrow, task and model-specific reasoning and lack comprehensive verification. While they employ formal methodologies, they fail to acknowledge foundational process distinctions such as events versus states, event subsumption, or point versus interval-based semantics. While these methods provide benefits like structured vocabularies [13,14] and more meaningful data access [15], they fail to capture fundamental ontological commitments needed for robust and reusable process understanding.

To address these limitations, we introduce a framework for formalizing process knowledge applied throughout the process mining lifecycle to enable transparency and replicability. This is accomplished through what we call "process meaning patterns": first-order logic ontology patterns that correspond to common verification and inference steps in the process mining lifecycle.

The process meaning pattern framework (PMP) demonstrated in this paper is a step towards a more meaningful and practical approach to understanding processes, which can be understood in terms of two goals. Firstly, the goal of ontology-aware process mining: to provide a means to formalize process knowledge applied throughout the process mining lifecycle to enable transparency and replicability. Secondly, the goal of data-aware process ontology: to have process ontologies informed by both the available data and the needs of the domain, striking a balance between accurately reflecting complex realities and providing useful abstractions for specific applications. In this paper, we focus on our contributions towards the former goal.

The contributions of this paper include: (1) a presentation of the process meaning pattern framework as a means to implement ontology-aware process mining. (2) Providing formal specifications of process meaning patterns based on real-world applications and data. (3) Providing a ready-to-use python-datalog implementation of the framework demonstrated on a real-world application from Volvo IT Belgium.

The paper is structured as follows: Sect. 2 expands on the motivation through real-world process mining challenges, focusing on process data quality and semantic heterogeneity. Section 3 critiques key literature from business process management, process mining, database theory, and formal ontology. Section 4 introduces the process meaning pattern methodology, including first-order logic axiomitizations and demonstrates this methodology using our Python and Datalog-based implementation applied to real enterprise data and use cases from Volvo. Finally, Sect. 5 situates the paper within process mining and knowledge representation research, including notable future research directions.

2 Process Meaning Challenges

Process mining transforms event logs and enterprise data into actionable process insights. However, it relies on implicit domain assumptions while lacking formal or standardized means of documentation, at the cost of reproducibility and verifiability. These challenges are further amplified by data heterogeneity and the lack of standardized handling of process data quality issues.

Process Data Quality Issues refer to event logs being noisy and incomplete, requiring pre-processing steps including data cleaning, filtering, and transformations. These steps are frequently viewed as secondary to core tasks such as process discovery, which constructs process models from event logs, or conformance checking, which compares logs' observed behaviour against expected models [1]. Recent deep learning approaches treat noise implicitly [16], while more traditional techniques assume clean data [17], obscuring valuable process knowledge embedded in these pre-processing tasks. Process mining is an iterative process: insights from discovery may prompt adjustments to data handling, and these pre-processing steps should be explicitly modelled rather than handled in an ad-hoc manner. Capturing event relationships, state transitions, and ordering constraints formally would make these steps modular and reusable, enhancing adaptability across mining tasks.

Recent work introduces imperfection patterns as a taxonomy of process mining data quality issues, outlining their impact, detection, and resolution strategies [3]. While potentially useful in practice, these patterns lack formal grounding and overlook the role of process knowledge in resolving data issues. For instance, the unanchored event pattern covers issues where timestamp formats vary (e.g., MM/DD/YYYY vs. DD/MM/YYYY). This is a syntactic issue simply requiring format standardization. By contrast, the *homonymous label* pattern describes issues where identical event labels refer to distinct behaviours, an issue of semantic heterogeneity. For example, a hospital log with several consecutive Triage Assessment events may in-fact represent distinct actions, such as an initial assessment and a follow-up. Resolving this issue requires reconciliation of process constraints and behaviours, not simply renaming labels or changing formats. Without formalizing such process knowledge, ad-hoc fixes obscure important semantics, reducing transparency and reuse. A structured, ontology-driven approach to data quality handling would improve consistency, verifiability, and integration within the process mining pipeline.

Semantic Heterogeneity of Event Logs is a result of the central artifact of process mining, the event log, being commonly recorded in the extensible event stream (XES) format [18]. XES is intentionally flexible, defining events as "an atomic granule of activity that has been observed,". This flexibility introduces semantic heterogeneity when representing diverse process entities such as status values, activities, resources, and occurrences. The Business Process Intelligence Challenges (BPIC)[1], widely used as process mining benchmarks [4,19], highlight these challenges. For instance, BPIC 2012 poses the challenge to "Identify which decisions have greater influence on the process flow". Addressing this challenge requires understanding complex process constructs including decisions, control flow, and causality - concepts not explicitly defined in event logs. Moreover,

---

[1] tf-pm.org/competitions-awards/bpi-challenge/.

event data quality and granularity vary significantly; across ten BPIC datasets, counts of unique activity labels range from 4 to 624[2].

The XES lifecycle extension classifies events with "lifecycle transitions" such as start, complete, schedule, or abort. However, real-world logs often lack explicit lifecycle tagging, making interpretation unclear. For example, BPIC 2017 explicitly tags events like "Complete Application" with start and complete transitions, while BPIC 2013's activity labels model status values (ex: "Accepted," "Queued"), with lifecycle transitions refining them (ex: "Awaiting Assignment"). Without standardization or ontological grounding, event log interpretation remains a key challenge, impacting consistency and integration of knowledge within the process mining pipeline.

3 Related Work

Many recent works have attempted to formalize process mining, including through ontologies. However, they face limitations in reasoning, verification, and scope, often tailoring logic to specific applications while neglecting fundamental ontological commitments. Moreover, existing process ontologies lack a practical grounding and are thus difficult to integrate into process mining pipelines.

Event Knowledge Graphs (EKGs) [20] extend process mining by representing multiple process perspectives based on objects involved in the processes like purchase orders or shipments. This is accomplished using labelled property graphs constructed via Cypher queries[3]. Unlike formal knowledge graph methods relying on RML/R2RML mappings [21], these mappings remain implicit in query structures, limiting transparency and reuse. Later refinements [11] introduce JSON-based "semantic headers" for mapping, but this weak encoding raises concerns for logical consistency. EKGs also assume perfect data quality, reinforcing the need for formalized data quality handling within process mining knowledge representations.

DECLARE [8] is a declarative process modeling approach that specifies constraints (e.g., "A happens at most once," "C always follows B") using linear temporal logic (LTL) rather than imperative sequences like $A \implies B \implies C$, with a defined constraint vocabulary and diagrammatic representations.

DECLARE's use of LTL has key semantic limitations, as it models only sequential relationships within a single case and treats all constraint operands as homogeneous event entities. These limitations prevent explicit reconciliation of semantic heterogeneity issues, and leave no room to model important heterogeneous process relationships like participation or event-state relationships. Additionally, semantic integration is virtually impossible as DECLARE's point-based semantics cannot be mapped to interval-based semantics, to name one issue.

---

[2] We provide our own quantitative analysis of these datasets at github.com/riley-momo/ProcessMeaningPatternsPython/tree/main/notebooks.

[3] https://neo4j.com/docs/cypher-manual/current/introduction/.

Data-aware DECLARE [9] extends DECLARE through an expanded metric first-order temporal logic (MFOTL) to incorporate data conditions via "event payloads", enhancing expressiveness but still reducing complex entities to homogeneous attribute-value pairs. This limitation, along with DECLARE's implicit process ontology, prevents fundamental inference and consistency checks, such as ensuring activities do not end before they begin. Other similar data-aware approaches are summarized in [10], where research in database theory dynamics for process analysis is presented. These approaches similarly use temporal logic and automated planning (AP) tools. However, our ontological critiques similarly hold, as the use of domain-specific schemata to validate logs fails to formally model fundamental ontological distinctions of processes. The reduction of diverse process and domain entities into homogeneous attribute-value pairs results in an inflexible and opaque semantics.

Ontology-Based Data Access (OBDA) for Process Mining [22], notably represented by onprom [15], maps data to an RDF graph to enable SPARQL queries for entity-centric retrieval. However, it suffers from poor expressiveness and neglects data pre-processing and quality issues. RDF graphs in OBDA support only subsumption, disjunction, and cardinality, which limits its ability to represent complex process concepts like temporal relationships. Moreover, OBDA relies on annotations and ad-hoc reasoning, limiting its formal reasoning capabilities.

Semantic Web Ontologies have been applied in process mining, with relatively high expressiveness, but still facing key limitations. For example, the Event Ontology (EVO) [14] serves as a taxonomy for event types like start or monitoring events, while BPMO [13] can annotate events according to control flow structures of common process models. Despite using OWL ontologies, these approaches rely on simple subsumption and domain/range axioms, making them more suitable for semantic tagging than for robust reasoning in diverse process mining scenarios. Other semantic web ontologies do not directly cover process mining but focus on events. These ontologies, however, are similarly limited by poor expressiveness, as the findings of Katsumi and Gruninger [23] indicates. These findings demonstrate that several prominent event ontologies are unable to handle basic competency questions of processes. Similarly, Benevides et al. [24] found that semantic web ontologies are unsuitable for representing foundational ontologies of process in their attempt to adapt the expressive UFO-B process ontology into a semantic web format.

Upper Ontologies model general, domain-independent knowledge that inevitably includes processes. Notable upper ontologies in this category include DOLCE [5] and UFO [6]. Being expressive and domain-independent ontologies, they consequentially carry conceptual bloat and redundancy of knowledge for process mining. For instance, to model the single simple event of a man running, DOLCE requires relevant perdurants, accomplishments, agentive physical objects, temporal qualities, and temporal regions. These ontologies' attempt to provide a general ontological reality leads to a confusing data structure for specific grounded applications like process mining. Additionally, these upper ontologies lack any tooling, methodology, or guidance for integration with domain datasets.

The Process Specification Language (PSL) [7] is a first-order logic ontology designed for the exchange of process knowledge among disparate information systems. While it is recognized as an ISO standard (ISO 18629[4]), its application for event logs is unclear due to their inherent semantic heterogeneity. PSL also lacks a means of integration with domain data or a practical way to handle exceptional behaviours like activity attempts, compounded by (general) undecidability issues in first-order logic reasoning.

4 The Process Meaning Pattern Framework

The process meaning pattern (PMP) framework formalizes the ontological commitments of process mining through semi-automated integration of ontology reasoning with data analysis tooling. In this section, we present the architecture of the framework and formally specify key demonstrative patterns with first-order logic and a logic programming implementation[5].

4.1 Overview

The high-level architecture of PMP (Fig. 1) can be viewed through two layers: a **knowledge layer** and an **implementation layer**, representing the formally encoded components of the framework, and the supporting software to integrate those formal components, respectively.

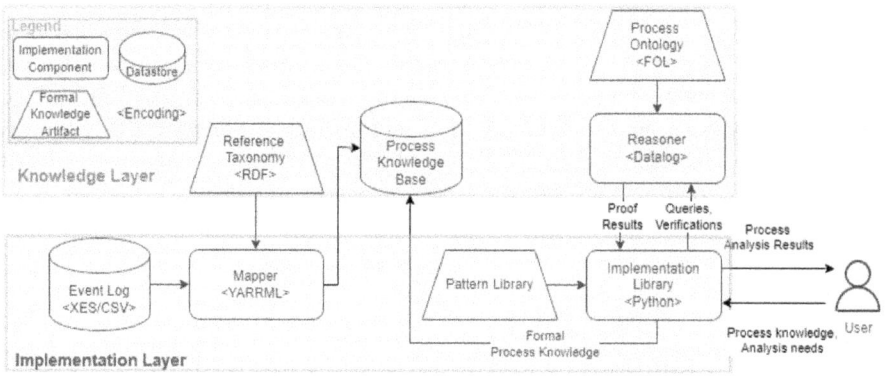

Fig. 1. Basic System Architecture for the PMP Framework

---

[4] https://www.iso.org/standard/35431.html.
[5] The full implementation of the framework is available at https://github.com/riley-momo/ProcessMeaningPatternsPython.

The end-user implementation is designed to mimic well-established libraries like pandas[6], with ontology-interpreted data structures rather than dataframes. Raw event log data is mapped into a set of simple RDF triples (e.g., hasRecordedTime(bakePizza, 12/03/24 10:15), case(baking03)), via a lightweight reference taxonomy (encoded via a YARRML file [25]), forming the initial process knowledge base.

The central construct of our framework is the **process meaning pattern**: specification of process statements, queries, and checks, like "What kinds of activities cause the status X to change?" or "X is a subactivity of Y". The patterns are specified via first-order logic sentence patterns in the language of the process ontology. While this is structurally similar to approaches like DECLARE [8], it has a richer semantics due to its reliance on an expressive process ontology. For the purpose of demonstrating the framework in this paper, we use a process ontology based on the process specification language (PSL) [7], though its aforementioned challenges will need to be addressed for a full implementation of the framework. By iteratively validating and extending the knowledge base through these patterns, PMP enables practitioners to formally capture assumptions, refine their analyses, and integrate provenance information directly into the process mining pipeline, rather than apply it in an ad-hoc manner.

Process Meaning patterns can be broadly categorized into three types: descriptors, facts, and queries; each of these is defined in table 1.

Table 1. Summary of Process Meaning Pattern Categories

Category	Description	Example
Descriptor	Specifies process behaviour using domain-specific vocabulary, quantified over ontology concepts	Gold members receive a 10% discount at checkout
Fact	States simple, quantifier-free knowledge about a process using domain-specific terms	Bob's loan was approved
Query	Validates or infers behavior by checking entailment of quantified statements over ontology concepts	What are all the instances of handoffs in loan approvals?

Consider an exemplary instance of a descriptor pattern: "When a fragile object is dropped, it breaks." - this can be expressed with the following first-order sentence in the language of the ontology:

$$(\forall o)\ occurrence\_of(o, drop) \wedge prior(fragile, o) \implies holds(broken, o)$$

This sentence is a specific example of a **state-based effect (SBE)** pattern, since it specifies how a state effects the result of an activity occurring. This

---
[6] https://pandas.pydata.org/.

process meaning pattern can be expressed through the meta-logical definition:

$$SBE(a, f_1, f_2) := (\forall o) \; occurrence\_of(o, a) \land prior(f_1, o) \implies holds(f_2, o)$$

In natural language, this states: "If an action a occurs and a state f_1 holds prior, then another state f_2 must hold afterwards."

4.2 Reasoning with Patterns

In addition to PMP enabling formal representations of process knowledge, use of a reasoner enables inference of additional process facts, retrieval of knowledge, and verification of process data. The positioning of the process ontology as a core element from which patterns are specified and data is mapped results in more focused coverage than other upper ontologies [5,6], while also distinguishing domain-independent knowledge from domain process knowledge, unlike other declarative approaches [8,9]. Additionally, reasoning through patterns presents an explicit connection between domain knowledge and process mining scenarios. For instance, the query pattern `handOffs(p)` is a unary query pattern, accepting a singular process variable, indicating this pattern may be instantiated without defining any process description or domain-level information. Conversely, a descriptor pattern like `stateBasedEffect(a, f1, f2)` accepts several arguments, indicating the knowledge necessary to express this rule. Since reasoning is framed in terms of the process ontology and a structured vocabulary of patterns, its complexity can be managed by restricting the expressiveness of allowed sentence patterns. This enables some control over decidability and tractability in process mining applications.

4.3 Implementation and Demonstration

To demonstrate our framework in action, we draw from a real-world case study from Volvo IT Belgium[7] with reference to specific python and datalog code.

One of the key challenges for Volvo IT Belgium (and for large enterprises in general) is understanding organizational dynamics within their processes. In this case, identifying teams and cases involved in "ping-pong" (also called multi-hop) behaviour, where several hand-offs occur while handling an IT incident. By applying PMP, we address this scenario by identifying and inferring instances of ping-pong behaviour directly from the dataset, demonstrating the power of our approach in capturing and reasoning about process dynamics. This demonstration will be presented in terms of the practitioners workflow.

Firstly, the event log data (modelled in CSV format) will be mapped into a process knowledge base containing a record of all the events pertaining to the IT incident management process. In this case, the mapping outputs to a datalog format, but our implementation also currently supports RDF, prover9, and CLIF formats.

---

[7] http://ais.win.tue.nl/bpi/2013/challenge.html.

Next, the pattern relevant to the current analysis needs would be selected, the ping-pong pattern in this case. In a future implementation, selection from a library of patterns would be integrated into the python library, but for now it must be manually entered in the datalog file. The ping-pong behaviour is defined by the following first-order sentence in the language of the ontology:

$$\forall (c)\, ping\_pong(c) \iff \exists (e_1, e_2, e_3, e_4, r_1, r_2)\, hand\_off(e_1, r_1, e_2, r_2, c) \wedge$$
$$hand\_off(e_3, r_3, e_4, r_4, c) \wedge (e_1 \neq e_3) \wedge (e_2 \neq e_4)$$

The ping_pong(c) pattern is true of a case c, and is defined in terms of the sub-pattern hand_off(e1, r1, e2, r2, c), which denotes a hand-off between the subsequent events e1 and e2 from resource r1 to resource r2 in the case c. Notice that this definition fits the query pattern as an existential conjunction of literals, and that ping pong is defined only in terms of the provided data; it does not require any background or domain-specific information to be applied.

When the query pattern is run, the reasoner will firstly check consistency of the pattern instance and knowledge base, and then return satisfying instances of the query, if any exist[8]. In this example, the datalog query pattern being matched is:

```
ping_pong(C) :-
    hand_off(R1, R2, E1, E2, C),
    hand_off(R3, R4, E3, E4, C),
    \+ (E1 = E3),
    \+ (E2 = E4).
```

The reasoner's output returns all cases with matching ping-pong behaviour, and will enumerate all proofs of this; this is equivalent to identifying all the possible pairs of hand-offs in each satisfying case. Given the proof output of the reasoner, practitioners are able to analyze each specific instance contributing to the ping-pong behaviour and examine them more closely.

Going beyond the ping-pong example, the intuitive and parametric definition of patterns enables complex queries to easily be expressed to better understand the process dynamics between resources. For example, "Which hand-off events involve the finance department taking over"? can simply be expressed through the query: $hand\_off(e, r_1, e_2, Finance, c)$. Additionally, given the positioning of this reasoning as part of a pre-defined vocabulary of patterns, as well as reasoning being carried out in datalog, it results in many computational benefits. Firstly, given that any pattern being checked will be evaluated through a datalog expression, it will be equivalent to a horn clause and thus be decidable. Secondly, it opens up new avenues for non-monotonic reasoning such as defaults or exceptions.

---

[8] In its current state datalog reasoning is semi-automated but future work will directly integrate the reasoner into the python environment for greater ease of use.

5 Conclusion and Future Work

The process meaning patterns framework advances ontology-aware process mining by bridging practical process mining analysis with formal ontology. Addressing challenges such as semantic heterogeneity, data quality, and reasoning tractability, it enhances the transparency, interpretability, and verification of process mining analyses. Through formal axiomitization, pragmatic implementation, and real-world applications, it demonstrates both theoretical rigour and practical utility.

Future work includes expanding the pattern library, refining the implementation including direct integration of datalog reasoning and pattern selection with python, and applying the framework as a benchmark of other process ontologies' practical utility. By grounding process mining in a structured yet practical framework, this work supports more consistent, reusable, and insightful analyses, benefiting both research and industry.

References

1. van der Aalst, W., et al.: Process mining manifesto. In: Daniel, F., Barkaoui, K., Dustdar, S. (eds.) BPM 2011. LNBIP, vol. 99, pp. 169–194. Springer, Heidelberg (2012). https://doi.org/10.1007/978-3-642-28108-2_19
2. van der Aalst, W.M.P.: Process mining: a 360 degree overview. In: Process Mining Handbook, pp. 3–34. Springer, Heidelberg (2022)
3. Suriadi, S., Andrews, R., ter Hofstede, A.H., Wynn, M.T.: Event log imperfection patterns for process mining: towards a systematic approach to cleaning event logs. Inf. Syst. **64**, 132–150 (2017)
4. Lopes, I.F., Ferreira, D.R.: A survey of process mining competitions: the bpi challenges 2011-2018. In: Di Francescomarino, C., Dijkman, R., Zdun, U. (eds.) BPM 2019. LNBIP, vol. 362, pp. 263–274. Springer, Cham (2019). https://doi.org/10.1007/978-3-030-37453-2_22
5. Borgo, S., et al.: Dolce: a descriptive ontology for linguistic and cognitive engineering. Appl. Ontol. **17**(1), 45–69 (2022)
6. Guizzardi, G., et al.: UFO: unified foundational ontology. Appl. Ontol. **17**(1), 167–210 (2022)
7. Grüninger, M.: Ontology of the process specification language. In: Handbook on Ontologies, pp. 575–592. Springer, Heidelberg (2004). https://doi.org/10.1007/978-3-540-24750-0_29
8. Di Ciccio, C., Montali, M., et al.: Declarative process specifications: reasoning, discovery, monitoring. Process Mining Handb. **448**, 108–152 (2022)
9. Bergami, G., Maggi, F.M., Marrella, A., Montali, M.: Aligning data-aware declarative process models and event logs. In: Polyvyanyy, A., Wynn, M.T., Van Looy, A., Reichert, M. (eds.) BPM 2021. LNCS, vol. 12875, pp. 235–251. Springer, Cham (2021). https://doi.org/10.1007/978-3-030-85469-0_16
10. Calvanese, D., De Giacomo, G., Montali, M.: Foundations of data-aware process analysis: a database theory perspective. In: Proceedings of the 32nd ACM SIGMOD-SIGACT-SIGAI Symposium on Principles of Database Systems, pp. 1–12 (2013)

11. Swevels, A., Fahland, D., Montali, M.: Implementing object-centric event data models in event knowledge graphs. In: International Conference on Process Mining, pp. 431–443. Springer, Heidelberg (2023). https://doi.org/10.1007/978-3-031-56107-8_33
12. Khayatbashi, S., Hartig, O., Jalali, A.: Transforming event knowledge graph to object-centric event logs: a comparative study for multi-dimensional process analysis. In: International Conference on Conceptual Modeling, pp. 220–238. Springer, Heidelberg (2023). https://doi.org/10.1007/978-3-031-47262-6_12
13. Norton, B., Cabral, L., Nitzsche, J.: Ontology-based translation of business process models. In: 2009 Fourth International Conference on Internet and Web Applications and Services, pp. 481–486. IEEE (2009)
14. Pedrinaci, C., Domingue, J.: Towards an ontology for process monitoring and mining. In: CEUR Workshop Proceedings, vol. 251, pp. 76–87 (2007)
15. Calvanese, D., Kalayci, T.E., Montali, M., Tinella, S.: Ontology-based data access for extracting event logs from legacy data: the onprom tool and methodology. In: Abramowicz, W. (ed.) BIS 2017. LNBIP, vol. 288, pp. 220–236. Springer, Cham (2017). https://doi.org/10.1007/978-3-319-59336-4_16
16. Bukhsh, Z.A., Saeed, A., Dijkman, R.M.: Processtransformer: predictive business process monitoring with transformer network. arXiv preprint arXiv:2104.00721 (2021)
17. Dunzer, S., Stierle, M., Matzner, M., Baier, S.: Conformance checking: a state-of-the-art literature review. In: Proceedings of the 11th International Conference on Subject-Oriented Business Process Management, pp. 1–10 (2019)
18. Gunther, C.W., Verbeek, H.M.W.: Xes-standard definition (2014)
19. Thiede, M., Fuerstenau, D., Barquet, A.: How is process mining technology used by organizations? A systematic literature review of empirical studies. Bus. Process. Manag. J. **24**(4), 900–922 (2018)
20. Fahland, D.: Process mining over multiple behavioral dimensions with event knowledge graphs. In: Process Mining Handbook, pp. 274–319. Springer, Heidelberg (2022). https://doi.org/10.1007/978-3-031-08848-3_9
21. Dimou, A., et al.: Rml: a generic language for integrated rdf mappings of heterogeneous data. In: Ldow, vol. 1184 (2014)
22. Xiao, G., et al.: Ontology-based data access: a survey. In: International Joint Conferences on Artificial Intelligence (2018)
23. Katsumi, M., Grüninger, M.: Using PSL to extend and evaluate event ontologies. In: Bassiliades, N., Gottlob, G., Sadri, F., Paschke, A., Roman, D. (eds.) RuleML 2015. LNCS, vol. 9202, pp. 225–240. Springer, Cham (2015). https://doi.org/10.1007/978-3-319-21542-6_15
24. Benevides, A.B., Bourguet, J.-R., Guizzardi, G., Peñaloza, R., Almeida, J.: Representing a reference foundational ontology of events in sroiq. Appl. Ontol. **14**(3), 293–334 (2019)
25. Heyvaert, P., De Meester, B., Dimou, A., Verborgh, R.: Declarative rules for linked data generation at your fingertips! In: Gangemi, A., et al. (eds.) ESWC 2018. LNCS, vol. 11155, pp. 213–217. Springer, Cham (2018). https://doi.org/10.1007/978-3-319-98192-5_40

How to Use a FEM Model as a Basis for Strategic-Level Risk Analysis

Toomas Saarsen[1](✉) and Ilia Bider[1,2]

[1] University of Tartu, Tartu, Estonia
toomas.saarsen@ut.ee
[2] DSV - Stockholm University, Stockholm, Sweden
ilia@dsv.su.se

Abstract. Risk analysis is a crucial aspect of ensuring smooth day-to-day operations. When an organization actively engages in process modeling and analysis, it gives a well-structured daily workflow. However, it is impossible to eliminate the potential impact of risks – risks cannot always be fully mitigated, and their realization can lead to negative consequences. Additionally, new risks (threats) emerge daily, which we have not encountered before. Therefore, it is essential to conduct risk analysis to minimize the negative impact of these threats. There are numerous methods and structures at the operational level to support risk management. This article will examine how the Fractal Enterprise Model could be used as a base for analyzing complex risk scenarios at the strategic level.

Keywords: Business Process Model · Fractal Enterprise Model · Risk Analysis · Risk Scenario

1 Introduction

Risk analysis [1] is carried out in various fields, for example, assessing product risks in finance [2] or evaluating production-related risks in organizations [3]. There are different stages involved in conducting a risk analysis [4].

Risk identification involves defining potential threats and assessing their likelihood of occurrence. Domain expertise and knowledge are crucial here, as they allow for identifying risks. Practical experiences, such as customer complaints or adverse outcomes of the process, highlights the risks and gives essential input for risk identification. In such cases, the risk has often materialized, and we address the risk analysis to prevent recurrence in the future.

Each identified risk must be evaluated during **risk assessment** regarding its probability of occurrence and potential consequences. A wide variety of methods can be used for this assessment, ranging from simple structures to complex computational models. In practice, various tables are used, which are easy to manage and provide the basic structure for analysis.

Risk treatment follows after assessing risks - decisions must be made on mitigating them and minimizing their negative impact on the organization. Possible measures involve avoiding risks, reducing their impact, or transferring them. All of these measures typically involve additional tasks for the organization. If the probability of a specific risk is high, the tasks associated with mitigating risks may become an integral part of the daily process.

Given the large and complex amount of information handled in the various stages of risk analysis, its structured description and handling are crucial [5]. When conducting a risk analysis, knowledge of the system within which the analysis is carried out is primarily needed [6]. System knowledge is essential in all three previously mentioned stages. It is necessary to find **the assets** that will be directly affected by the realization of the risk. When assessing the risk, it is necessary to analyze the **negative impact** of the risk on the system. When addressing the risk, it is necessary to find actions and means to **reduce the likelihood of the risk and its negative impact**. The risk analyst must possess knowledge of the system. Still, in the case of large systems, a structured description is also essential from which missing (forgotten) details can be found if necessary, and distant connections between different system elements can be seen, which a person cannot do with a large number of facts and connections [7].

Another important piece of information relates directly to the risk and its associated information - which assets the risk may affect, the probability of the risk, and the negative impact [8]. When describing a risk, the most popular format is a table - a structure for listing and analyzing different parameters of the risks. These tables provide a general structure and make it relatively easy to order them according to various risk parameters. For example, one table used in risk analysis in industrial enterprises is FMEA [9]. FMEA table contains a list of risks in one dimension (usually rows) and risk-related details and parameters in the other dimension (usually columns). Such a table is helpful in systematizing risks and performing simple calculations, which aids the risk assessor in prioritization.

In the final stage of risk analysis – Risk Treatment – solutions must be found to reduce the probability of the risk-related threat and its resulting impact [10]. The solution typically involves changes to the system: additional tasks are planned, critical components are duplicated, and additional resources or backups are created. It is beneficial for the risk analyst to first analyze how the realization of the risk would affect the system in its current state, plan the necessary changes, and explore the system's functioning in the event of the risk's realization after the changes have been made.

This article presents a case study demonstrating the application of a FEM [12] model in strategic risk analysis. We investigate the specific information that can be incorporated into a FEM model to support risk analysts and how this model facilitates the risk analysis process.

2 Background

Data necessary for risk analysis can be found in the process model [11], which describes how the company operates. This model illustrates how the institution functions and how critical assets are utilized. The **strategic level** (general view) provides an overview of the general processes. It can be likened to a table of contents, indicating the processes within the company. Porter's value chain [11] is often used here to present a complete list of processes and highlight their relationships. The **operative level** focuses on detailed activities performed by actors daily. BPMN diagrams [11] are frequently used here to provide a more detailed view of the sequence of activities and assets needed. At a **detailed level**, it is aimed at devices and automation.

Many companies use process diagrams (BPMN) to conduct risk analysis on operational level; process diagrams provide a structural picture of the sequence of activities and the use of assets. Unlike the BPMN diagram, the Porter's value chain on the strategic level does not include the context of the assets required. A Fractal Enterprise Model (FEM) model [12] provides us with such a high-level structure of asset usage. The FEM model brings together various types of assets at a general level and shows the relations that arise through the use of these assets in general processes. In the FEM diagram (see Fig. 1), assets are shown with rectangles and processes with ovals. The relations are shown in the FEM model with a line between processes and assets.

The FEM model structure is similar in structure to a process diagram but at a more general level (strategic level). If detailed structured views aid process analysts in granular risk analysis on an operative level, could similar structures support risk assessments on a strategic level? Since the companies we have previously collaborated with did not use general structured diagram views for risk analysis, we became interested in testing the FEM model in conducting risk analysis on a strategic level. Based on the above, the question arose for investigation:

- *how to use the FEM model, as a general structured form of a process, for conducting strategic-level risk analysis?*

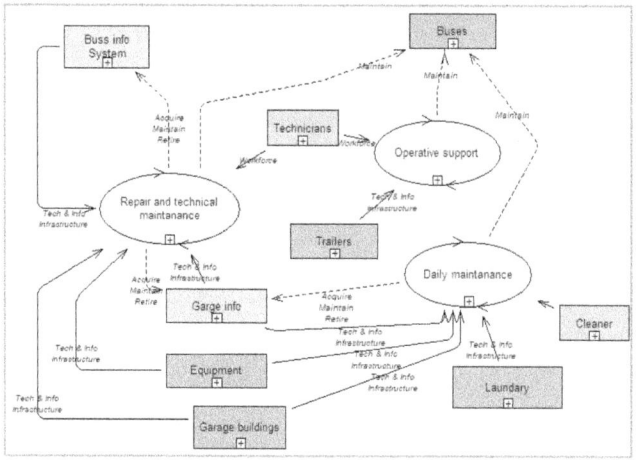

Fig. 1. Description of a bus company support unit using FEM.

We selected two companies with experience in organization modeling: a **transportation company** providing both scheduled and on-demand bus services and a **logistics company** handling freight between foreign countries and local sales companies. We generated FEM models based on the existing process models. The FEM model was a foundation for constructing and examining risk scenarios. The diagrams and examples presented in the article are taken from the study of the transportation company.

3 Structures for Strategic Risk Analysis

In operative level risk analysis, often, two structures are used in parallel: an FMEA table for listing different risks and process diagrams for analyzing risks on a timeline.

Assets that are influenced during risk realization could be highlighted in the process diagram; it gives suitable bases for analyzing impacts through infrastructure (e.g., problems in one software system can propagate to other software systems that are directly synchronized and share data), and the flow of effects in the process chain (if the output of one task is essential for the execution of a subsequent task, the following tasks is also likely to be affected).

In a more general (strategic level) risk analysis, a similar structure can be helpful for the risk analyst - a timeline through general processes in the context of assets around the processes. In the following, we will look at how to move from detailed process diagrams to a more general FEM model, which could be used for strategic risk analysis.

3.1 Basic Information to Go from Process Model for FEM

We use data from the process model for building the FEM model for risk analysis on a strategic level. We assume that the organization has a process model:

- Process model covers the entire organization at a high level with value chain(s).
- In essential areas, process diagrams (e.g., BPMN or EPC [13]) have been created at the detail level, where the critical context of tasks is determined.
- Processes are described using a modeling tool that relies on a database (repository), from which it is possible to automatically retrieve meaningful connections between processes and the assets associated with the tasks of processes.

To ensure a comprehensive model encompassing the entire organization, we should incorporate the following views from the process model into the FEM model: **a value chain** offers a broad overview of all organizational processes; **various views of assets** provide a structured representation of the elements used in detailed process diagrams, such as data, personnel, and equipment.

Most process models already include structured views for different types of assets. They could be structured, for example, as organizational structure (a hierarchical representation of the organization's actors), data relationships (a graphical depiction of how different data sets are related), or inventory (a structured list of other important assets (for example, materials or equipment)). If there is no general view about specific types of assets, then such a view could be created based on a detailed list of elements from the process model as a report - a query from the process model repository to get a list of one type of assets. These elements from the list should be aggregated into more general view elements - a general view of assets.

Overall Structure of FEM

Processes systematizes various infrastructure elements (assets) in the FEM model. We take the first or second-level value chains from the existing process model (depending on the level of detail) and combine them into a general value chain covering the entire organization, showing approximately ten general processes in sequential order. We took a second-level value chain (over 20 processes) and presented it through ten processes (see Fig. 2).

As a second step, we extract infrastructure elements from the process model - essential assets for performing a single task. The assets associated with tasks in the process diagram are too detailed for a general model and need to be generalized and aggregated into more general elements. Due to the high number of detailed assets in large process models, it is common practice to structure them on separate diagrams to simplify management and improve overview, for example, a separate diagram of general data elements or the organization's structure. If there is no map for some of the assets, then such a map could be created based on a detailed list of elements from the process model as a report (query from the repository and aggregation into a more general view).

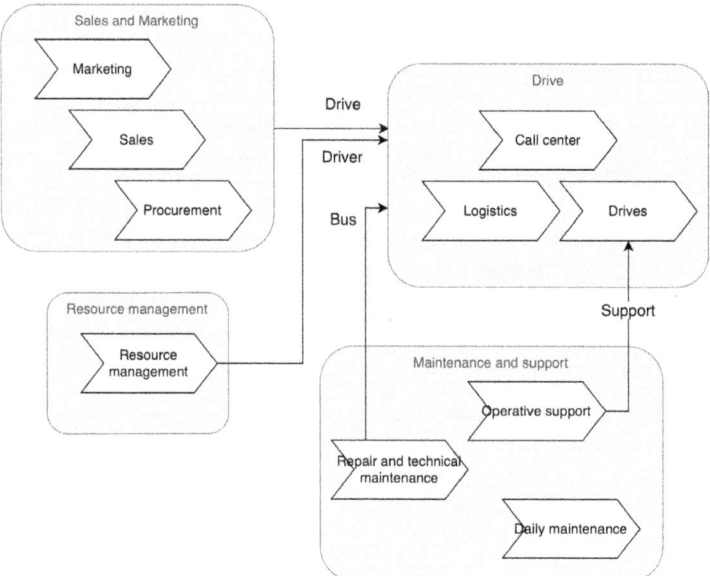

Fig. 2. Value chain for risk analysis

We imported general assets into the FEM model:

- **Actor-assets** - actors implementing processes and using previously highlighted assets during the process (marked in green in diagrams). We derived actors from the organizational hierarchy.
- **Data-assets** - software and different devices for data management (marked in red in diagrams). We sourced data from a generalized logical data model, augmented with a server overview specifying the software, hardware, and data managed by each server.

- Another type of asset - different **resources** (devices, buildings, or materials) needed to implement daily processes (marked in blue in diagrams). The assets that are not material but have value for the business could be included here (for example, services, fame, or feedback). We call them **virtual assets** and mark them in orange in diagrams). In a separate table, we listed these resources (not data or actors) and aggregated them into more general elements (assets).

3.2 FEM Model Based on Value-Chain and Asset-Maps

Based on the previous information, we put together a FEM model and used it in the creation and analysis of risk scenarios. In creating the FEM model, we based the layout on the value chain, as the company's employees were accustomed to this layout, and it is easier to follow when conducting a strategic risk analysis. The original FEM model contained a total of over 50 elements, which were difficult to track and use in risk analysis. We suggested breaking down the model into smaller diagrams categorized by asset type: a diagram focused on data assets (see Fig. 3, where the FEM model used in the context of risk scenario is presented) and a diagram focused on resource assets (see Fig. 4, where the FEM model in the context of risk scenario is presented). Both views included actors, highlighting who performed the processes and used the associated assets. This preference was justified because risk analysis scenarios often concentrate on particular asset types. For instance, when assessing IT-related risks, the focus is on the data, and everyday resources like materials or vehicles are less relevant.

4 Case Study

To answer the question highlighted in Sect. 2 (Background), we tested two steps previously described: (1) extracted the needed data from the process model and built the FEM model; (2) based on the FEM model structure, we created various risk scenarios and examined how risk analysts use the diagrams in different phases of risk analysis.

4.1 Case Study Setting

As we already mentioned in Sect. 3.2, we tested the FEM model in two organizations:

- A transportation company in Estonia (belongs to the top three) offering various bus transportation services. The company has around 500 employees.
- A logistics company in Estonia with around 150 employees focused primarily on goods storage and logistics planning. The company is part of an international corporation.

Companies use modeling software for business process management. The bus company created its process model a few years ago and used it in practice. The second company process model was under development and not tested actively. The first company assets in the model were already structured and generalized into the bigger general units. The second company generalized its assets onto the maps based on a list of detailed resources during FEM model building.

4.2 Case Study Protocol

The companies were already using detailed risk analysis daily at a detailed level - daily reclamations, problems, and new ideas. Our case study focused on strategic risk analysis, for example, questions arising during the strategic-level budgeting or global risks affecting the organization.

In the course of our case study, we constructed diverse risk scenarios underpinned by a FEM diagram. We investigated their utility in risk identification, analysis, and modification.

Risk Scenario for the Risk Treatment

When defining a risk scenario, it is crucial to determine its starting point (which asset is initially affected) and analyze how it evolves and which other organizational elements will be impacted. We highlighted the entry point of the risk realization in the diagram - assets directly influenced. Next, we marked the influence flow via timeline and infrastructure.

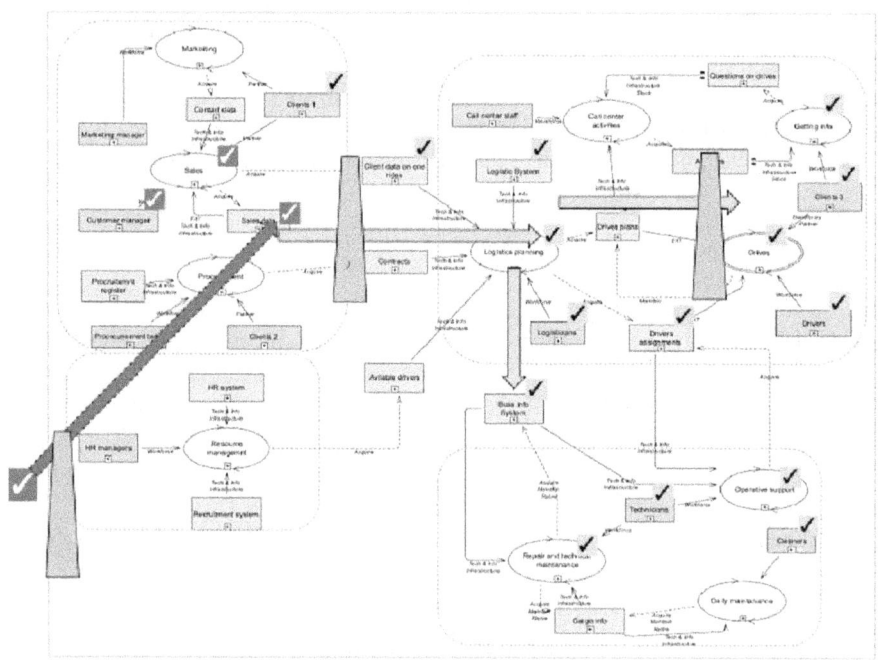

Fig. 3. Risk mitigation.

Figure 3 highlights the negative impact on the organization caused by a data management failure. We marked the direct impact (which assets and processes are directly affected) in red and the track of how the negative impact spreads through the relationships to other assets in yellow. This scenario was a good exercise in playing out at which point in the process, the negative impact of the risk could be stopped or reduced (see brown towers (trapeziums) in Fig. 3).

This view identifies the information systems potentially impacted should the risk occur. The diagram illustrates that the impact's progression can be stopped at various stages to avoid disrupting the following process steps. However, every additional security measure or system redundancy incurs a specific cost (investment). This diagram proved helpful in determining the necessary investments to mitigate the negative impact at different points - each tower comes at a price.

Improving system security is important here to mitigate external risks. We planned system duplication and replacement for critical data assets to ensure sustainable operation - we designed all these changes on a detail-level process diagram.

The Risk Scenario for the Calculating the Price of the Risk
A central question in risk analysis is its cost - if a risk materializes, what is its monetary impact, or in other words, what is the cost to the organization? The FEM model provided a solid framework for identifying affected assets, assigning values to them, and finally calculating a total value (see Fig. 4).

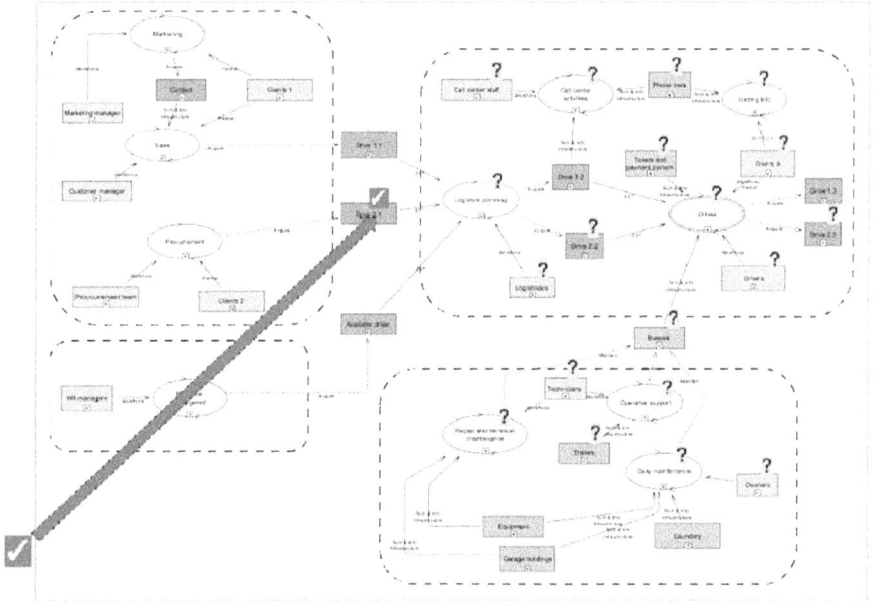

Fig. 4. The price of the risk.

We marked all resource assets affected after risk realization in red and all influenced assets in yellow. We estimated the negative value of change (cost) for every asset. The sum of the calculated costs of assets gives a general estimation of the total negative impact after risk materialization.

Based on Fig. 4, a rough risk estimate has been calculated – a rough price range in the event of the risk materializing. When moving through the diagram, it was relatively easy to assess the local context around red assets (elements around the initially affected assets), and the estimates were also likely accurate. Each subsequent step on this complex graph was more challenging to assess and likely had a more significant margin of error. According to the risk analysts, the value of the FEM diagram was primarily in obtaining a complete picture of what would be affected in the organization. The estimate (range) of the risk price obtained as a result of the calculation provided more of a sense of the financial magnitude/size of the risk.

The diagram (see Fig. 4) highlights assets influenced by indirect relationships. The principal value provided by such a diagram is completeness - we can obtain elements for the asset list that we would not initially consider in the calculation.

4.3 Findings

Risk analysts, experienced in using process diagrams for granular risk assessments, affirmed the value of the FEM structure in strategic risk analysis and highlighted three critical aspects in defining risk scenarios on the FEM model.

- FEM diagrams helped track how the realization of a specific threat could damage a particular asset, the processes that would be directly impacted, and **how this impact could spread** through the organization (the overall process) over time.
- Another essential value that FEM diagrams provided was the assessment of the value of a risk. In most cases, the realization of a risk does not only disrupt a single asset and a single process. Such a comprehensive network reflecting resources helped highlight **all the assets** disrupted due to the risk **in the calculation**.
- The important topic was finding the **"starting point"** – at which point(s) the risk materializes in the value chain.

A positive aspect highlighted was that different analysts had different views of how a specific risk would materialize, and such a structured model provided everyone with the same picture.

General comment on the diagram, already known from previous experience - diagrams must be simple. Managers and specialists involved in the risk analysis should focus on something other than deciphering diagrams; they should focus on the risk assessment and treatment. Risk scenarios involve numerous elements and relations that are too complex for a human to grasp fully. Dividing essential elements into different views (separate diagrams) simplifies reading complex relationships and identifying key points/content on the diagrams.

4.4 Limitations

There were only two organizations involved in the case study. It was a good start for testing the FEM model as a structure for risk scenarios. The following case studies are already planned and probably will give additional and more detailed feedback about the usability of the FEM model in the context of risk analysis.

5 Conclusion

FEM structure supports risk analysis at the general strategic level. It provides valuable information and structure for the risk identification and analysis phases. Different risk scenarios could be built on the FEM model, and analyze different aspects of risk realization could occur.

We will continue to apply the FEM model in other companies to investigate how to improve its usability and assess the quality of the FEM model within the context of such risk scenarios.

Acknowledgment. The work of the second author was partly supported by the Estonian Research Council (grant PRG1226).

References

1. Aven, T.: Risk Analysis. John Wiley & Sons, Hoboken (2015)
2. Yermoshenko, A.M.: Scenario analysis as a risk management instrument in cooperation of insurers and commercial banks. Actu. Prob. Econ. **100**, 88–96 (2009)
3. Silvestri, A., De Felice, F., Petrillo, A.: Multi-criteria risk analysis to improve safety in manufacturing systems. Int. J. Prod. Res. **50**(17), 4806–4821 (2012)
4. Muehlen, M.Z., Ho, D.T.Y.: Risk management in the BPM lifecycle. In: International Conference on Business Process Management, pp. 454–466. Springer, Heidelberg (2005)
5. Haimes, Y.Y.: Risk modeling of interdependent complex systems of systems: theory and practice. Risk Anal. **38**(1), 84–98 (2018)
6. Bolger, F., Wright, G.: Use of expert knowledge to anticipate the future: Issues, analysis and directions. Int. J. Forecast. **33**(1), 230–243 (2017)
7. Alhawari, S., Karadsheh, L., Talet, A.N., Mansour, E.: Knowledge-based risk management framework for information technology project. Int. J. Inf. Manag. **32**(1), 50–65 (2012)
8. Janes, H., Pepe, M.S., Gu, W.: Assessing the value of risk predictions by using risk stratification tables. Ann. Intern. Med. **149**(10), 751–760 (2008)
9. Sharma, K.D., Srivastava, S.: Failure mode and effect analysis (FMEA) implementation: a literature review. J. Adv. Res. Aeronaut. Space Sci. **5**(1–2), 1–17 (2018)
10. Brown, I., Steen, A., Foreman, J.: Risk management in corporate governance: a review and proposal. Corp. Gover. Int. Rev. **17**(5), 546–558 (2009)
11. Dumas, M., Rosa, L.M., Mendling, J., Reijers, A.H.: Fundamentals of Business Process Management. Springer-Verlag, Heidelberg (2018)
12. Bider, I., Perjons, E., Elias, M., Johannesson, P.: A fractal enterprise model and its application for business development. Softw. Syst. Model. **16**, 663–689 (2017)
13. Nizioł, M., Wisniewski, P., Kluza, K., Ligęza, A.: Characteristic and comparison of UML, BPMN and EPC based on process models of a training company. Ann. Comput. Sci. Inf. Syst. **26**, 193–200 (2021)

Modelling Neural Network Models

Nadia Daoudi[1]( ), Ivan Alfonso[1], and Jordi Cabot[1,2]

[1] Luxembourg Institute of Science and Technology, Esch-sur-Alzette, Luxembourg
{nadia.daoudi,ivan.alfonso,jordi.cabot}@list.lu
[2] University of Luxembourg, Esch-sur-Alzette, Luxembourg

Abstract. Neural networks are now a key element of many complex software systems, typically known as AI-enhanced systems or smart systems. Model-driven engineering (MDE) approaches are often used to model and (semi)automatically generate such complex systems. Nevertheless, for MDE to be used in the creation of smart systems, it is essential to be able to model the implementation of neural networks (NNs) as part of the rest of the system. Unfortunately, modelling support for NNs is rather limited. In particular, there is a lack of complete and expressive neural network metamodels that could facilitate NN development. This poses significant challenges as it creates a disconnect between the development of neural networks and the other components of the system. Overall, this paper introduces a metamodel and its concrete syntax to support neural network design, providing a foundation for future research in smart system development.

Keywords: Metamodel · Model · Neural networks · Grammar

1 Introduction

Neural networks have become a prevalent component in the design of modern software. With the growing demands for more intelligent and efficient software, neural networks empower systems to enhance their functionality and responsiveness. The abundance of data has also significantly increased the reliance on neural networks to perform complex tasks, driving advancements in areas such as predictive analytics and recommender systems.

Offering versatility across different components of a system's architecture, neural networks can be leveraged for both front-end and back-end tasks. On the front-end side, they provide functionalities to support tasks involving direct interactions with users. On the back-end, NNs can be integrated to support critical tasks involving advanced data processing and analysis. These functionalities can be accomplished using various types of neural networks, including convolutional neural networks (CNNs) and recurrent neural networks (RNNs).

Neural networks are generally developed separately from the other components of the system due to the limited availability of modeling approaches specifically designed for neural networks. Indeed, NN metamodels are a missing piece in smart software development. In the literature, some studies have attempted to propose metamodels for neural networks [2,5,12]. However, they suffer from limitations, such as being very specific to particular types of neural networks (e.g., a metamodel for convolutional neural networks) or failing to comprehensively address basic neural network types and the different components involved in their development, such as datasets and training setup.

To fill this gap, we contribute a metamodel and its concrete syntax for modelling neural networks. Our NN metamodel covers fundamental components necessary for NN development along with a textual notation for the concrete syntax. Specifically, the metamodel encompasses the elements needed for architecture design, dataset loading, training, and performance evaluation. Our work aims to facilitate neural networks development as it would support future research in enabling NNs to be modelled as seamlessly as any other system component. All the artefacts from our work are available in an open-source repository[1].

The rest of this paper is organised as follows: Sect. 2 presents basic concepts related to NN development. In Sect. 3 we describe our metamodel and the concrete syntax. Section 4 presents an evaluation of our metamodel. Finally, we review related research in Sect. 5 and conclude the paper in Sect. 6.

2 Background

Neural networks are designed to recognise and identify patterns present in a given dataset in order to perform complex tasks such as classification or regression. The main building blocks of neural networks are layers which represent computational units that perform mathematical operations to transform the input data. Neural networks can also potentially be composed of tensorOps (i.e., short for tensorOperations) and sub-neural networks. TensorOps are low-level operations applied to tensors (i.e., multidimensional arrays of data) within and between layers [6]. Such operations include addition, reshaping, and multiplication to manipulate the data. In some complex architectures, neural networks can include other sub-neural networks as building blocks to create a unified model.

Layers in neural networks can be of different types, such as fully connected layers, convolutional layers and recurrent layers. Fully connected layers perform matrix operations that connect each neuron in the previous layer to every neuron in the subsequent layer. Convolutional layers apply convolutional filters to detect patterns in data, making them particularly effective for tasks such as image processing. Recurrent layers are designed for sequential data processing by maintaining internal memories. Since they can capture temporal dependencies and patterns in the input data, this type of neural networks is effective for tasks such as time series forecasting and natural language understanding. Convolutional and recurrent layers have various sub-types designed to perform specific tasks and handle different data structures [7,14]. We show in Fig. 1 an example of an NN architecture taken from a TensorFlow tutorial[2]. The neural network

---
[1] https://github.com/BESSER-PEARL/BESSER.
[2] https://www.tensorflow.org/tutorials/images/cnn.

consists of various types of layers for image classification, including Conv2D, MaxPool2D, Flatten, and LinearLayer (i.e., fully connected layer).

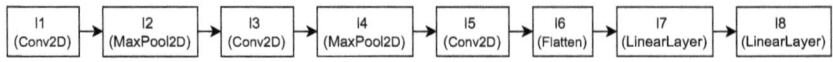

Fig. 1. Example representing an NN model composed of eight layers.

The order of layers in neural network architectures can be of two types:

- Sequential architecture: Layers are structured in a linear, straightforward sequence. The output of a layer is passed to its subsequent layer sequentially.
- Non-sequential architecture: In this type of architecture, layers are not arranged in a simple linear order. Instead, connections can skip layers, allowing a layer to receive input from earlier, non-adjacent layers. This creates a structure where the flow of information does not follow a strict sequence.

To help neural networks learn complex patterns in the data, activation functions can be used to introduce non-linearity to the transformations applied by the layers. After a neural network architecture is designed, hyperparameters need to be defined to guide the learning process and optimise the model's performance. Such hyperparameters can include the optimiser [11], the loss function [19] and the learning rate [4]. Besides, a well curated training dataset also needs to be made available to the neural network so it can learn meaningful patterns. The effectiveness of an NN model is heavily dependent on the quality, diversity, and size of the dataset used in the training. After the neural network is trained, a test dataset is used to assess its learning and validate its performance.

3 NN Modelling Language

In this section, we present our NN metamodel and its concrete syntax.

3.1 The NN Metamodel

We provide the class diagram of our NN metamodel in Fig. 2. The key concepts in the NN metamodel are represented using meta-classes and their associations. Our design was heavily inspired by the two popular deep learning frameworks PyTorch[3] and TensorFlow[4]. Specifically, we compared concepts from the two frameworks to come up with a metamodel design that is general to represent neural network components and allows the definition of both sequential and non-sequential architectures. This approach ensures that the metamodel remains versatile and adaptable across different contexts within neural networks development. In the following, we present the main metaclasses that define our metamodel and some of their important attributes.

---

[3] https://pytorch.org.
[4] https://tensorflow.org.

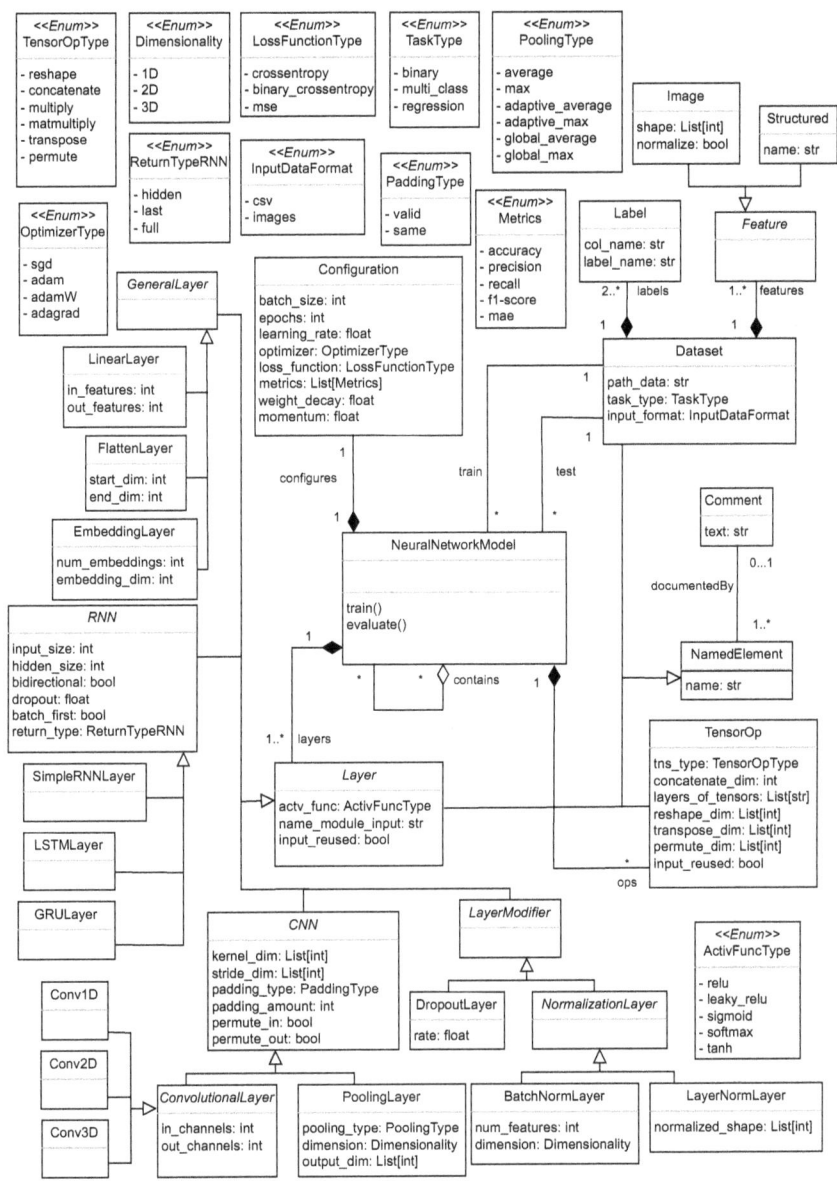

Fig. 2. The NN metamodel.

NeuralNetworkModel. At its core, the metamodel contains the *NeuralNetworkModel* metaclass that comprises properties and behaviours of an NN model. A *NeuralNetworkModel* can contain three types of components, referred to as *modules*: (sub)-*NeuralNetworkModel*s, *Layer*s and *TensorOp*s. Specifically, a *NeuralNetworkModel* can incorporate another *NeuralNetworkModel* as a component (*contains* relationship in the metamodel). This is common in NNs, where a model can include other models as part of its architecture. The other two modules (i.e., *Layer* and *TensorOp*) are described in the following paragraphs.

Layer. A *Layer* is an abstract metaclass that encapsulates properties of NN layers. It defines the activation function (i.e., *actv_func* attribute) that is detailed through an enumeration listing the possible types of activation function, such as relu, sigmoid, and softmax. As it is the case for all the enumerations in the metamodel, the list of activation functions can be extended as needed. In our metamodel, a layer can be of four types, represented by the following abstract metaclasses: *GeneralLayer*, *LayerModifier*, *CNN*, and *RNN*.

TensorOp. This metaclass is used to represent an operation or function applied to tensors. In our metamodel, six types of tensorOps are defined in the *TensorOpType* enumeration. The *type* attribute of the *TensorOp* metaclass gets its value from this enumeration. Beside *type*, the *TensorOp* metaclass includes additional attributes that serve as parameters for the tensor operations.

Configuration. This metaclass contains attributes representing several hyperparameters essential for training neural networks, and an attribute *metrics* to store the metrics used for evaluating the model's performance.

Dataset. Neural networks rely on datasets to learn and generalise based on examples. In our metamodel, it is represented using the metaclass *Dataset* that can serve either as the training or the test dataset.

Comment. Neural networks components are generally adopted for different reasons (i.e., a sub-NN has shown promising results in another architecture, a type of layer is well suited to the task, a layer has given better results based on trial and error, ...). To keep track of these design decisions, we use the *Comment* metaclass that adds relevant documentation to any *NamedElement*. Note that several concepts inherit from *NamedElement* in our metamodel, facilitating the reuse and sharing of NN models.

3.2 Concrete Syntax

Neural networks (NNs) can be described using various notations. In this work, we propose a textual notation for their definition, supported by a grammar we developed to instantiate the concepts of our metamodel.

Listing 1.1 presents an excerpt of our grammar, developed using ANTLR[5]. The primary rule, *neuralNetwork*, allows for specifying a name (ID), layers, sub-neural networks, tensor operations, modules to represent the sequential architecture of the NN, parameters (defined in the subsequent parser rule in line 12),

---
[5] https://www.antlr.org/.

along with datasets for training and testing. We employ specific operators in ANTLR to enforce cardinalities in our metamodel: the '+' operator indicates one or more occurrences, '?' denotes zero or one, and '*' represents zero or more.

A key rule in the grammar is *layer* (lines 21–23), where each layer must have a defined name (ID) and a set of parameters, which vary depending on the layer type. For every layer type, we define a dedicated parser rule. For instance, the *linear* rule (lines 31–34) contains general layer parameters (defined by the *layerParams* rule on line 25), such as the activation function, along with specific parameters for linear layers, such as *in_features* and *out_features*.

Our language supports the definition of NNs using the *modules* rule (line 41), which enables the user to specify the order of layers, sub-neural networks, and tensor operations for the NN.

Finally, we define in lines 43–48 the rule for specifying the training dataset, including the name (ID), *path_data*, *task_type*, and other parameters as defined in the metamodel. The complete grammar is available in the project repository.

Listing 1.1. DSL grammar excerpt

```
grammar NN;

neuralNetwork   : ID ':'
                  'layers' ':' layer+
                  ('sub_nn' ':' sub_nn+)*
                  ('tensor_ops' ':' tensorOp+)*
                  'modules' ':' modules
                  'config' ':' configuration
                  'trainingDataset' ':' dataset
                  'testingDataset' ':' dataset ;

configuration   : 'batch_size' '=' INT
                  'epochs' '=' INT
                  'learning_rate' '=' DOUBLE
                  'optimiser' '=' STRING
                  'metrics' '=' strList
                  'loss_function' '=' lossFunction
                  'weight_decay' '=' DOUBLE ;
                  'momentum' '=' DOUBLE ;

layer           : '-' ID ':'
                  (generalLayer | rnn
                  | cnn | layerModifier) ;

layerParams     : ('actv_func' '=' activityFuncType)?
                  ('name_layer_input' '=' STRING)?
                  ('input_reused' '=' BOOL)? ;

generalLayer    : linear | flatten ;

linear          : 'type' '=' 'Linear'
                  layerParams
                  'in_features' '=' INT
                  'out_features' '=' INT ;

flatten         : 'type' '=' 'Flatten'
                  layerParams
                  ('start_dim' '=' INT)?
                  ('end_dim' '=' INT)? ;

modules         : ('-' ID)+ ;

dataset         : 'name' '=' ID
```

```
'path_data'    '='  STRING
'task_type'    '='  taskType
'input_format' '='  inputFormat
'image'        '='  intList
'labels'       '='  '{' label ',' label (',' label)? '}' ;

//Other rules omitted for brevity purposes ...
```

Listing 1.2 provides a textual example of the neural network (NN) model shown in Fig. 1. The model definition begins by specifying the NN's name (my_model). Next, the layers are defined (lines 2–19) outlining three layers (l1, l2, and l3), with l1 and l3 being 2D Convolutional layers, and l2 as a Pooling layer. Then, the modules definition (lines 21–22) specifies the order of the layers. Finally, hyperparameters are defined (lines 23–29), such as the "adam" optimiser. The full textual model can be accessed in the project repository.

Listing 1.2. NN textual model example

```
my_model:
    layers:
      - l1:
          type=Conv2D
          actv_func=relu
          in_channels=3
          out_channels=32
          kernel_dim=[3, 3]
      - l2:
          type=Pooling
          pooling_type=max
          dimension=2D
          kernel_dim=[2, 2]
      - l3:
          type=Conv2D
          actv_func=relu
          in_channels=32
          out_channels=64
          kernel_dim=[3, 3]
      //Other layer definitions omitted for brevity.
    modules:
      - l1 - l2 - l3 - l4 - l5 - l6 - l7 - l8
    config:
        batch_size=32
        epochs=10
        learning_rate=0.001
        optimiser="adam"
        metrics=["f1-score"]
        loss_function=crossentropy
//Dataset definitions ommited for brevity.
```

4 Evaluation

In this section, we aim to evaluate the expressiveness of our metamodel. Specifically, we select four state-of-the-art NN architectures from the literature and rely on our approach to model them:

- AlexNet [8], which is composed of various layers including Conv2D, Pooling, Dropout and Linear layers. It contains two sub-NNs: one to extract the features and the other to conduct the classification.

- VGG16 [15], which incorporates Conv2D, Pooling, Dropout and Linear layers. It is composed of two sub-NNs for feature extraction and classification.
- LSTM [21], which is an RNN-based neural network since it contains LSTM layers in addition to Linear, Dropout and Embedding layers.
- CNN-RNN [20], which includes a mix of CNN and RNN layers such as Conv1D, GRU, Linear, Dropout and Embedding layers. It also uses the concatenate tensorOp and implements a non-sequential architecture.

Our attempt to model these NNs helped us refine our metamodel to include some concepts that we overlooked in our initial design. For instance, the LSTM model uses a bidirectional LSTM layer that processes data in both forward and backward directions to capture information from past and future contexts within a sequence. Our metamodel was missing this "bidirectional" concept and was then updated after our attempt to model the LSTM neural network. We have successfully modelled the four NNs using our approach, which demonstrate its effectiveness in expressing state-of-the-art neural networks. The four modelled neural networks are available in our repository.

5 Related Work

Neural networks are becoming an integral component of intelligent systems, driving advancements in various domains. An NN metamodel has been proposed [5] to improve the verification and versioning of AI components. While addressing some NN concepts, this metamodel does not cover fundamental layer types, limiting its overall applicability. Similarly, an NN markup language has been proposed [12]; however, it remains basic as it defines networks in terms of connected processing units, and does not address the specific structures and connections found in complex architectures like CNNs and RNNs. Another NN metamodel [2] has been proposed, only covering multi-layer perceptrons. An UML-based design flow has been proposed to streamline NN deployment [17]. Nonetheless, this design has limited coverage, as it does not encompass basic layer types.

To support the design of convolutional neural networks (CNNs) in autonomous vehicle perception (AVP), a metamodel [13] has been proposed. While the metamodel has similarities to our work, its application domain is narrow as it only covers CNNs which makes it very specific and not appropriate to model other well-known NN types such as RNNs. An NN metamodel [9] has been proposed to incorporate predictive tasks into manufacturing. It represents NNs using basic neuron and edge abstractions, omitting other fundamental NN types and potentially limiting its versatility. To automate AI-based software development processes [10], a metamodel has been proposed addressing the procedural aspects of AI development. However, it does not specifically target ML or NN concepts such as algorithms and hyperparameters.

Several research works have also proposed domain-specific languages (DSLs) to enhance computational efficiency in ML. DeepDSL [22] enhances the portability and efficiency of deep learning models. TensorFlow Eager [1] provides an interactive model to improve development flexibility. Diesel [3] optimises GPU performance for linear algebra and NN computations. FastML [18] facilitates the parallelisation of ML algorithms on mult-icore processors. OptiML [16] simplifies and optimises ML tasks execution through implicit parallelism.

While the above works develop metamodels or DSLs for neural networks, their applicability and focus is limited. With our work, we aim to propose a comprehensive metamodel that covers the fundamental types of NN layers along with the other concepts needed for NN training and evaluation.

6 Conclusion

To support smart component development, we designed an NN metamodel that covers fundamental concepts to neural network development. Our metamodel captures the elements necessary for architecture design (layers, tensorOps, sub-NNs), dataset loading, training and performance evaluation. Additionally, we contributed a textual notation for the concrete syntax to facilitate NN definition and conducted an evaluation to assess the expressiveness of our approach. All the artefacts from our work are available in an open-source repository.

As further work, we will work on the development of a useful ecosystem of tools around this metamodel such as importers, code-generators and model transformations and even model-based migration processes from/to neural networks implemented with different libraries.

Acknowledgments. This project is supported by the Luxembourg National Research Fund (FNR) PEARL program, grant agreement 16544475.

References

1. Agrawal, A., et al.: Tensorflow eager: a multi-stage, python-embedded dsl for machine learning. In: Proceedings of Machine Learning and Systems, vol. 1, pp. 178–189 (2019)
2. Al-Azzoni, I.: Model driven approach for neural networks. In: International Conference on Intelligent Data Science Technologies and Applications, pp. 87–94 (2020)
3. Elango, V., Rubin, N., Ravishankar, M., Sandanagobalane, H., Grover, V.: Diesel: DSL for linear algebra and neural net computations on GPUs. In: Proceedings of the 2nd ACM SIGPLAN International Workshop on Machine Learning and Programming Languages, New York, NY, USA, pp. 42–51 (2018)
4. He, F., Liu, T., Tao, D.: Control batch size and learning rate to generalize well: theoretical and empirical evidence. In: Advances in Neural Information Processing Systems, vol. 32. Curran Associates, Inc. (2019)
5. van den Heuvel, W.-J., Tamburri, D.A.: Model-driven ML-Ops for intelligent enterprise applications: vision, approaches and challenges. In: Shishkov, B. (ed.) BMSD 2020. LNBIP, vol. 391, pp. 169–181. Springer, Cham (2020). https://doi.org/10.1007/978-3-030-52306-0_11
6. Ji, Y., Wang, Q., Li, X., Liu, J.: A survey on tensor techniques and applications in machine learning. IEEE Access **7**, 162950–162990 (2019)

7. Ketkar, N., Moolayil, J.: Convolutional Neural Networks, pp. 197–242. Apress, Berkeley (2021)
8. Krizhevsky, A., Sutskever, I., Hinton, G.E.: ImageNet classification with deep convolutional neural networks. In: Advances in Neural Information Processing Systems, vol. 25. Curran Associates, Inc. (2012)
9. Lechevalier, D., Hudak, S., Ak, R., Lee, Y.T., Foufou, S.: A neural network metamodel and its application for manufacturing. In: 2015 IEEE International Conference on Big Data (Big Data), pp. 1428–1435 (2015)
10. Morales, S., Clarisó, R., Cabot, J.: Towards a DSL for AI engineering process modeling. In: International Conference on Product-Focused Software Process Improvement, pp. 53–60. Springer, Heidelberg (2022). https://doi.org/10.1007/978-3-031-21388-5_4
11. Nwankpa, C.: Advances in optimisation algorithms and techniques for deep learning. Adv. Sci. Technol. Eng. Syst. J. **5**(5), 563–577 (2020)
12. Rubtsov, D., Butakov, S.: A unified format for trained neural network description. In: IJCNN'01. International Joint Conference on Neural Networks. Proceedings (Cat. No.01CH37222), vol. 4, pp. 2367–2372 (2001)
13. Safdar, A., Azam, F., Anwar, M.W., Akram, U., Rasheed, Y.: MoDLF: a model-driven deep learning framework for autonomous vehicle perception (AVP). In: the International Conference on Model Driven Engineering Languages and Systems, MODELS '22, New York, NY, USA, pp. 187–198 (2022)
14. Schmidt, R.M.: Recurrent neural networks (rnns): a gentle introduction and overview. arXiv preprint arXiv:1912.05911 (2019)
15. Simonyan, K., Zisserman, A.: Very deep convolutional networks for large-scale image recognition. arXiv preprint arXiv:1409.1556 (2014)
16. Sujeeth, A.K., et al.: OptiML: an implicitly parallel domain-specific language for machine learning. In: Proceedings of the 28th International Conference on International Conference on Machine Learning, ICML'11, pp. 609–616 (2011)
17. Suárez, D., Posadas, H., Fernández, V.: UML-based design flow for systems with neural networks. In: 2023 38th Conference on Design of Circuits and Integrated Systems, pp. 1–6 (2023)
18. Tsopgny, N.G.N., Nguélé, T.M., Kouakam, E.: DSL for parallelizing machine learning algorithms on multicore architecture. In: CARI 2022 (2022)
19. Wang, Q., Ma, Y., Zhao, K., Tian, Y.: A comprehensive survey of loss functions in machine learning. Ann. Data Sci., 1–26 (2020)
20. Wang, X., Jiang, W., Luo, Z.: Combination of convolutional and recurrent neural network for sentiment analysis of short texts. In: Proceedings of the 26th International Conference on Computational Linguistics, pp. 2428–2437 (2016)
21. Yang, S., Yu, X., Zhou, Y.: LSTM and GRU neural network performance comparison study: taking yelp review dataset as an example. In: International Workshop on Electronic Communication and Artificial Intelligence, pp. 98–101 (2020)
22. Zhao, T., Huang, X.: Design and implementation of DeepDSL: a DSL for deep learning. Comput. Lang. Syst. Struct. **54**, 39–70 (2018)

Quantifying the Magnitude of Violation: Predictive Compliance Monitoring Approaches

Qian Chen[1(✉)], Stefanie Rinderle-Ma[1], and Lijie Wen[2]

[1] Technical University of Munich, Garching, Germany
{qian.chen,stefanie.rinderle-ma}@tum.de
[2] Tsinghua University, Beijing, China
wenlj@tsinghua.edu.cn

Abstract. Most existing process compliance monitoring approaches detect compliance violations in an ex post manner. While predicate prediction methods address this by forecasting compliance violations, they offer only binary outcomes without quantifying how significantly an ongoing process instance deviates from the desired state. Measuring the magnitude of violations would provide organizations with deeper insights into operational performance, supporting informed decision-making to mitigate the risk of non-compliance. Thus, we propose two predictive compliance monitoring methods: the first transforms the binary classification into a hybrid classification-regression task, and the second leverages multi-task learning to simultaneously predict compliance status and quantify the magnitude of violation. We focus on temporal constraints as they are prevalent across domains such as healthcare. Evaluations on synthetic and real-world event logs demonstrate that our approaches enable violation quantification while maintaining comparable compliance prediction performance achieved by state-of-the-art approaches.

Keywords: Predicate prediction · Outcome-oriented predictive process monitoring · Temporal constraints · Magnitude of Violation

1 Introduction

Process compliance is a pivotal component in information systems research [1] and focuses on ensuring that processes adhere to compliance constraints. Compliance constraints may stem from regulatory documents, codes of practice, and business rules or contracts [8], might change frequently (e.g., financial regulations change every 12 minutes[1]), and non-compliance might result in the risk of financial loss and even criminal penalties [6].

Several surveys on challenges in compliance management in (process-aware) information systems exist, e.g., [1,13,14,18]. A first dimension of challenges arises

---
[1] https://thefinanser.com/2017/01/bank-regulations-change-every-12-minutes.

from the process life cycle phase when compliance verification is supposed to take place, i.e., design time, runtime and ex post. Despite the importance of design and ex post compliance checking, compliance monitoring enables organizations to detect and manage possible violations during executions [13]. However, only few existing approaches target the *prediction* of violations. Another challenge is the insufficient *ability to quantify the degree of compliance*, defined as a compliance monitoring functionality (CMF10) in [13]. Existing *outcome-oriented predictive process monitoring* and *predicate prediction* approaches [7,15,21,23] provide binary yes/no notions of compliance prediction at runtime, e.g., will an order be delivered on time (desired outcome) or not (undesired outcome). Yet, they do not quantify the extent of deviation from the desired state. Differentiating non-compliance levels of running cases can support an organization in making informed decisions to reduce or mitigate the risk of non-compliance.

To address these challenges, we propose two predictive compliance monitoring (PCM) [18] approaches: a hybrid method that reformulates binary classification into a regression task; and a multi-task learning (MTL) method that explicitly predicts compliance status and magnitude of violations, simultaneously. We focus on temporal violations due to their importance in domains like logistics and aviation [3], and the universal availability of timestamps in event logs. Evaluation on synthetic and real-world logs confirms that our approaches, MTL in particular, can effectively quantify the magnitude of violations with compliance prediction performance comparable to state-of-the-art methods. The extended version of this paper can be found in [4].

The paper is structured as follows. Section 2 presents a running example. The proposed PCM approaches are introduced in Sect. 3 and evaluated in Sect. 4. Related work is discussed in Sect. 5, followed by the conclusion in Sect. 6.

2 Problem Statement

We illustrate the problem based on an Order-to-Cash (O2C) process (cf. Fig. 1) as the running example. The process is triggered when a purchase order (PO) is received. An associated temporal constraint φ_t is introduced into the process model that imposes a maximum temporal distance of 24 hours between tasks ship goods and confirm order.

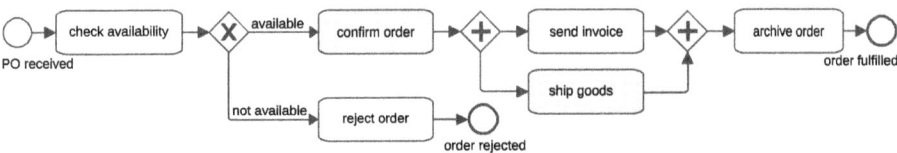

Fig. 1. Order-to-Cash Process Model

To predict compliance states of running cases w.r.t. φ_t, existing predicate prediction approaches [7,15,23] classify them either into satisfaction or violation via a binary classifier. Consider the event log generated from the running example in Fig. 2. Predicate prediction approaches label every case in the log according to φ_t and train the binary classifier based on the labeled event log to predict which of the running case in evolving event streams will and will not violate φ_t.

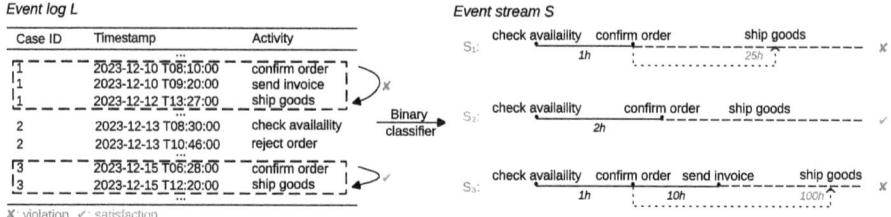

Fig. 2. Predicate Prediction

By contrast, this work extends the binary compliance prediction with the ability to quantify the magnitude of violation. Consider streams s_1 and s_3 in Fig. 2 and assume a predicted duration of 25 and 100 hours between tasks confirm order and ship goods for s_1 and for s_3, respectively. Existing approaches would classify both as non-compliant, but s_3 exhibits a more significant extent of deviation. Quantifying the magnitude of violation supports risk-informed decision-making to reduce the likelihood or severity of process faults [5]. For example, potentially non-compliant cases with a "mild degree" of violation such as s_1 can be addressed first to reduce the number of cases with negative outcomes, as they are typically easier and less costly to resolve. On the other hand, severe violations like s_3 may be prioritized for intervention to mitigate their impact.

3 Approach

We propose hybrid and MTL approaches to extend predicate prediction methods (cf. Fig. 2) for magnitude quantification. The main distinctions in between lie in how we label cases and how to train the model.

3.1 Predicate Prediction

Case Labeling is conducted for each case in the log based on the corresponding temporal constraint. An event log $L := \{\sigma_1, ..., \sigma_k\}$ comprises a set of traces σ_i where $\sigma_i := \langle e_1, e_2, ..., e_n \rangle$ is defined as a sequence of events e_j. An event $e_j := (c, a, t, d_1, ..., d_m)$ is defined as a tuple with case id c, activity a, timestamp t, and other attributes $d_1, .., d_m$. A temporal constraint φ_t specifies the control-flow and temporal requirements each process execution must comply with. The control-flow patterns confine the occurrences or ordering of activities, while time patterns impose restrictions on the (e.g., minimum or maximum) temporal distance in between. The labeling function for each case w.r.t. φ_t is defined as follows:

$$y_{\text{binary}}(\sigma) = \begin{cases} \text{deviant} \to 1 & \text{if } \varphi_t \text{ violated in } \sigma \\ \text{normal} \to 0 & \text{otherwise} \end{cases}$$

Specifically, a case is labeled as deviant (encoded as 1) if it violates φ_t; otherwise, it is labeled as normal (encoded as 0).

Model Training takes as input the encoded features and predicts their corresponding labels.

For a binary classification problem, we minimize the binary cross-entropy loss (BCE) between the true labels y_i (either 0 or 1) and the predicted values $\hat{y}_i$ (i.e., estimated probability within $[0, 1]$) for all (i.e., n) samples:

$$\mathcal{L}_{\text{BCE}} = -\sum_{i=1}^{n} y_i \log(\hat{y}_i) \tag{1}$$

3.2 Hybrid Approach

Case Labeling in this approach is different from y_{binary} (cf. Sect. 3.1). We follow up on the idea from [9] to assign numerical values for each case in the log as the magnitude of violation. In particular, the hybrid approach assigns a positive value to deviant cases reflecting the magnitude of violation, while normal cases are consistently labeled as 0. For deviant cases that breach temporal constraints in the control-flow dimension, we use the case duration (i.e., the time from the start to the end of the case) as the magnitude of violation to highlight the severity of activity violation. By doing so, it maintains consistency with the binary classification method for normal cases while offering a more nuanced evaluation of violations. The corresponding labeling function is defined as follows:

$$y_{\text{hybrid}}(\sigma) = \begin{cases} \text{magnitude of violation} & \text{if } \varphi_t \text{ violated in } \sigma \\ 0 & \text{otherwise} \end{cases}$$

Consider the running example (cf. Sect. 2) with φ_t imposing a maximum temporal restriction of 24 hours between confirm order and ship goods. The magnitude of violation for σ_1 is calculated as $53 - 24 = 29$ hours.

Model Training takes the same feature vectors as input, but predicts the magnitudes of violations instead of class labels. To enable compliance predictions, we convert the continuous predictions to binary values according to the hybrid labeling function y_{hybrid}. Specifically, cases are classified as deviant if their predicted values > 0, otherwise, they are considered normal. For the regression task, we aim to minimize the squared difference between the ground truth values y_i and predicted values $\hat{y}_i$ by using the mean squared error (MSE) as the loss function:

$$\mathcal{L}_{\text{MSE}} = \frac{1}{n} \sum_{i=1}^{n} (y_i - \hat{y}_i)^2 \tag{2}$$

3.3 Multi-task Learning

Case Labeling for MTL approach combines y_{binary} and y_{hybrid} to assign labels and the magnitudes of violations for each case, resulting in a labeled event log with both label and magnitude columns:

$$y_{\text{mtl}}(\sigma) = \begin{cases} y_{\text{binary}}(\sigma) \\ y_{\text{hybrid}}(\sigma) \end{cases}$$

Model Training in the MTL framework addresses both classification and regression tasks simultaneously. As illustrated in Fig. 3, both tasks share the same layers as they are closely related and likely share common representations. The compliance prediction task determines whether a case is deviant or normal. While for magnitude quantification, unlike the hybrid approach, it incorporates the output of the compliance prediction as an additional input. This integration is crucial because the magnitude of violation is only relevant for deviant cases, making the classification output a valuable signal for the regression task.

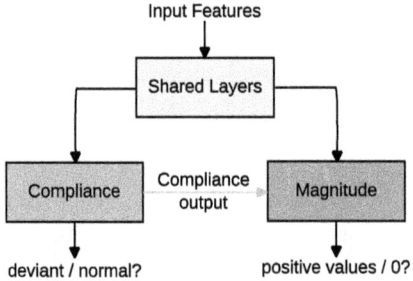

Fig. 3. Model Architecture for MTL

To train the MTL model, we minimize the total loss from compliance prediction (cf. Eq. 1) and magnitude quantification (cf. Eq. 2):

$$\mathcal{L} = \mathcal{L}_{\text{BCE}} + \mathcal{L}_{\text{MSE}} \tag{3}$$

4 Evaluation

We evaluate our proposed approaches by comparing them with predicate prediction methods (baseline). We employ two predictive models for all approaches, the XGBoost classifier [21] from ML and the Attention-based Bidirectional LSTM (Att-Bi-LSTM) [22] from DL. All experiments are conducted on synthetic (cf. Fig. 1) and real-life event logs[2] (cf. Table 1) with following temporal constraints:

---
[2] https://data.4tu.nl.

o2c_1: "Goods must be shipped no later than 24 hours after order confirmation."
sepsis_1: "Patients should be administered antibiotics within one hour after ER Sepsis Triage"
sepsis_2: "Lactic Acid measurements should be performed within three hours from ER Sepsis Triage."
bpic2012w_1: "A scheduled validation should start within 2 days."
bpic2012w_2: "The validation must be done in at most 20 minutes."

Table 1. Characteristics of Datasets

Dataset	#traces	#event classes	Length min.	avg.	max.	pos. class ratio	act. vio ratio
o2c_1	998	6	1	2.53	3	0.41	0.0
sepsis_1	782	15	1	7.19	61	0.63	0.2
sepsis_2	782	16	3	7.08	39	0.25	0.94
bpic2012w_1	9658	15	2	13.17	93	0.20	0.02
bpic2012w_2	9658	17	2	13.61	93	0.15	0.0

4.1 Experimental Setup

We remove incomplete cases and include all available attributes as well as time-related features (e.g., *weekday*) as input. XGBoost encodes categorical attributes via aggregation while Att-Bi-LSTM uses one-hot encoding. All traces are cut before the constraint-related events occur to allow early predictions. We keep the 80%-20% train-test-split ratio for model training and evaluation. Regarding model architectures, we only adapt the Att-Bi-LSTM model to the MTL framework as shown in Fig. 3 due to the flexibility of DL models. The shared layers include a bidirectional LSTM, an attention layer highlighting critical input features, global average pooling, a dense and a dropout layer. Hyperparameters are optimized using the Tree-structured Parzen Estimator (TPE) algorithm [2], with each configuration evaluated through 3-fold cross-validation.

To assess the performance, We use *area under the ROC curve (AUC)* as the metric for compliance prediction and *Mean Absolute Error (MAE)* for magnitude quantification. Since the predicate prediction approach does not provide magnitude predictions, we instead evaluate a trivial baseline by using the *average* of the actual values as predictions. The implementation is available at https://github.com/Qian915/predictive-compliance-monitoring.

4.2 Results

We report experimental results with XGBoost (cf. Table 2) and Att Bi-LSTM (cf. Table 3) to assess the performance of compliance prediction (AUC) and magnitude quantification (MAE in days).

XGBoost. For traditional ML models, such as XGBoost (cf. Table 2) in this study, our proposed approaches outperform the baseline in predicting the magnitude of violations, achieving lower MAE values across nearly all datasets. The only exception is the hybrid approach on the sepsis datasets, which may be attributed to the high proportion of deviant cases which violate the temporal constraints on the control-flow dimension (cf. activity violation ratio in Table 1). In other words, the constraint-relevant event does not even occur. For such severe violations, the case duration is used as the magnitude of violation, introducing a substantial number of outliers and resulting in a skewed data distribution. The hybrid approach, which focuses on regression tasks, is particularly sensitive to such imbalanced data distributions compared to the other two methods.

Regarding compliance prediction performance, the MTL approach achieves results comparable to the baseline. However, the hybrid approach shows a significant decline in compliance prediction performance for the sepsis_2 dataset, where the activity violation ratio is extremely high at 0.94 (cf. Table 1). By contrast, for the o2c_1 dataset, the hybrid approach demonstrates a notable improvement, with an AUC value of 0.70. This improvement can be attributed to the synthetic dataset containing only basic attributes but with many temporal features extracted from the timestamps, which enhance the predictive performance for a regression task.

Table 2. Comparison Performance for XGBoost

Dataset	Approach	Compliance AUC	Magnitude MAE
o2c_1	baseline	0.50	0.31
	hybrid	**0.70**	**0.21**
	multi-task	0.50	0.22
sepsis_1	baseline	**0.57**	6.06
	hybrid	0.51	8.41
	multi-task	**0.57**	**4.59**
sepsis_2	baseline	**0.87**	11.05
	hybrid	0.58	13.11
	multi-task	**0.87**	**9.83**
bpic2012w_1	baseline	**0.56**	1.42
	hybrid	0.51	1.28
	multi-task	0.55	**0.91**
bpic2012w_2	baseline	0.53	0.04
	hybrid	**0.55**	**0.02**
	multi-task	0.52	**0.02**

Att-Bi-LSTM. Table 3 summarizes the experimental results obtained with Att-Bi-LSTM. Similar to the magnitude prediction performance observed for XGBoost (cf. Table 2), both approaches outperform the baseline. Notably, when Att-Bi-LSTM is used as the predictive model, our proposed approaches—particularly the MTL approach—achieve lower MAE values compared to XGBoost. This mainly stems from the MTL framework, where the magnitude prediction task leverages patterns captured by the compliance prediction as an additional input.

The MTL approach slightly outperforms the baseline in compliance prediction. For the sepsis_2 dataset, the hybrid approach exhibits a similar decline in compliance prediction performance as observed with XGBoost (cf. Table 2), due to the high activity violation ratio. Meanwhile, for the bpic2012w dataset, the hybrid approach underperforms other methods. This is likely due to the extremely low positive case ratios of 0.20 and 0.15 (cf. Table 1), which result in imbalanced data distributions that negatively impact regression tasks.

Table 3. Comparison Performance for Att-Bi-LSTM

Dataset	Approach	Compliance AUC	Magnitude MAE
o2c_1	baseline	**0.50**	0.31
	hybrid	**0.50**	**0.23**
	multi-task	**0.50**	**0.23**
sepsis_1	baseline	**0.60**	6.06
	hybrid	0.58	8.17
	multi-task	0.56	**3.64**
sepsis_2	baseline	0.86	11.05
	hybrid	0.55	9.21
	multi-task	**0.87**	**8.67**
bpic2012w_1	baseline	0.57	1.42
	hybrid	0.50	1.34
	multi-task	**0.58**	**0.85**
bpic2012w_2	baseline	0.55	0.04
	hybrid	0.50	0.03
	multi-task	**0.56**	**0.02**

5 Related Work

We analyze related work on compliance prediction and compliance degree quantification w.r.t. a given set of constraints.

Compliance Prediction. [16] classify predictive process monitoring (PPM) approaches into process-aware and non-process-aware approaches, both employing regression and classification methods to predict attributes in the future, such as next activity, next time, or the outcome of the case. As part of PPM, outcome-oriented PPM [20,21] and predicate prediction [7,15] focus on compliance predictions of ongoing instances regarding a given set of constraints; the binary yes or no answer w.r.t. compliance is not sufficient for compliance management and requires the ability to quantify the degree of compliance.

Compliance Degree. [12] calculate a compliance distance to indicate the degree of match between the process model and control rules at design time. They define rules into four distinct classes according to the ideal semantics: ideal, sub-ideal, non-compliant and irrelevant states. [11] classify cases into full-, partial- and non-compliance and formulate a framework to detect and evaluate the degree of violations after process executions. However, the user-defined piece-wise mapping function cannot distinguish the extent of violations in a fine-grained manner. In [17], a compliance scale model is created to measure the degree of compliance of completed process instances. Still, the mechanisms of c-semirings to classify process instances into $<Good, Bad>$ or $<0, .5, 1>$ are not accurate enough. [19] define a set of key performance indicators (KPIs) for each compliance rule and map the evaluation values of them to a compliance level from -100 to 100 taking into account values of target, threshold and worst. However, existing approaches provide various metrics for compliance degrees in different phases of process life cycle, but none of them target compliance prediction.

6 Conclusion

This work proposes hybrid and MTL approaches to extend predicate prediction methods with the ability to quantify the magnitude of violation for an ongoing case. The evaluation demonstrates that our approaches—particularly the MTL method—outperform state-of-the-art methods in magnitude quantification while maintaining comparable performance in compliance prediction. The performance of both tasks improves further when DL models are adopted due to the flexibility in adapting model architectures. However, the performance of the hybrid approach is highly sensitive to data distribution, i.e., performing poorly on skewed datasets. This limitation is primarily due to the case labeling function applied to deviant cases, where the relevant event may not even occur. In future work, we aim to investigate methods for assigning the magnitudes of violation to deviant cases that breach temporal constraints in the control-flow dimension. Additionally, we plan to explore techniques to adapt the multi-task model architecture further, such as dynamically weighting each task's loss [10], to improve overall performance.

Acknowledgments. This work was funded by the Deutsche Forschungsgemeinschaft (DFG, German Research Foundation) – project number 277991500.

References

1. Abdullah, N.S., Sadiq, S.W., Indulska, M.: Emerging challenges in information systems research for regulatory compliance management. In: Advanced Information Systems Engineering, pp. 251–265 (2010)
2. Bergstra, J., Bardenet, R., Bengio, Y., Kégl, B.: Algorithms for hyper-parameter optimization. In: Advances in Neural Information Processing System, pp. 2546–2554 (2011)
3. Cheikhrouhou, S., Kallel, S., Guermouche, N., Jmaiel, M.: The temporal perspective in business process modeling: a survey and research challenges. SOCA **9**, 75–85 (2015)
4. Chen, Q., Rinderle-Ma, S., Wen, L.: Beyond yes or no: predictive compliance monitoring approaches for quantifying the magnitude of compliance violations. CoRR arxiv:2502.01141 (2025)
5. Conforti, R., de Leoni, M., Rosa, M.L., van der Aalst, W., ter Hofstede, A.: A recommendation system for predicting risks across multiple business process instances. Decis. Support Syst. **69**, 1–19 (2015)
6. Elgammal, A., Sebahi, S., Turetken, O., Hacid, M.S., Papazoglou, M., van den Heuvel, W.: Business process compliance management: an integrated proactive approach. In: Business Information Management Association, pp. 764–781 (2014)
7. Francescomarino, C.D., Dumas, M., Maggi, F.M., Teinemaa, I.: Clustering-based predictive process monitoring. IEEE Trans. Serv. Comput. **12**(6), 896–909 (2019)
8. Hashmi, M., Governatori, G., Lam, H.-P., Wynn, M.T.: Are we done with business process compliance: state of the art and challenges ahead. Knowl. Inf. Syst. **57**(1), 79–133 (2018). https://doi.org/10.1007/s10115-017-1142-1
9. Junior, J., Pinto, T., Morais, H.: Hybrid classification-regression metric for the prediction of constraint violations in distribution networks. Electric Power Syst. Res. **221**, 109401 (2023)
10. Kendall, A., Gal, Y., Cipolla, R.: Multi-task learning using uncertainty to weigh losses for scene geometry and semantics. In: Computer Vision and Pattern Recognition, pp. 7482–7491 (2018)
11. Lam, H., Hashmi, M., Kumar, A.: Towards a formal framework for partial compliance of business processes. In: AI Approaches to the Complexity of Legal Systems, pp. 90–105 (2020)
12. Lu, R., Sadiq, S.W., Governatori, G.: Measurement of compliance distance in business processes. Inf. Syst. Manag. **25**(4), 344–355 (2008)
13. Ly, L.T., Maggi, F.M., Montali, M., Rinderle-Ma, S., van der Aalst, W.: Compliance monitoring in business processes: functionalities, application, and tool-support. Inf. Syst. **54**, 209–234 (2015)
14. Ly, L.T., Rinderle-Ma, S., Göser, K., Dadam, P.: On enabling integrated process compliance with semantic constraints in process management systems - requirements, challenges, solutions. Inf. Syst. Front. **14**(2), 195–219 (2012)
15. Maggi, F.M., Francescomarino, C.D., Dumas, M., Ghidini, C.: Predictive monitoring of business processes. In: Advanced Information Systems Engineering, pp. 457–472 (2014)
16. Márquez-Chamorro, A.E., Resinas, M., Ruiz-Cortés, A.: Predictive monitoring of business processes: a survey. IEEE Trans. Serv. Comput. **11**(6), 962–977 (2018)
17. Morrison, E.D., Ghose, A., Koliadis, G.: Dealing with imprecise compliance requirements. In: EDOC Workshops, pp. 6–14 (2009)

18. Rinderle-Ma, S., Winter, K., Benzin, J.: Predictive compliance monitoring in process-aware information systems: state of the art, functionalities, research directions. Inf. Syst. **115**, 102210 (2023)
19. Shamsaei, A.: Indicator-based policy compliance of business processes. In: International Conference on Advanced Information Systems Engineering (2011)
20. Teinemaa, I., Dumas, M., Maggi, F.M., Francescomarino, C.D.: Predictive business process monitoring with structured and unstructured data. In: Business Process Management, pp. 401–417 (2016)
21. Teinemaa, I., Dumas, M., Rosa, M.L., Maggi, F.M.: Outcome-oriented predictive process monitoring: review and benchmark. ACM Trans. Knowl. Discov. Data **13**(2), 17:1–17:57 (2019)
22. Wang, J., Yu, D., Liu, C., Sun, X.: Outcome-oriented predictive process monitoring with attention-based bidirectional LSTM neural networks. In: Web Services, pp. 360–367. IEEE (2019)
23. Zhang, J., Liu, G.: Prediction of incompliance with business goals with business-related data and context data. IEEE Access **8**, 187008–187020 (2020)

Evaluating Programming Optimization Techniques in C and Python: Impact on Energy Consumption

Carlos Pulido[1](✉), Félix O. García[1], Mª Ángeles Moraga[1], Coral Calero[1], Miguel Baños-González[2], Jorge Cancho-Casado[2], and Javier Corral-García[2]

[1] Instituto de Tecnologías y Sistemas de Información, Camino de Moledores, s/n, 13005 Ciudad Real, Spain
{carlos.pulido,felix.garcia,mariaangeles.moraga,coral.calero}@uclm.es
[2] Fundación COMPUTAEX (Computación y Tecnologías Avanzadas de Extremadura), Cáceres, Spain
{miguel.banos,jorge.cancho,javier.corral}@computaex.es

Abstract. Software plays a critical role in modern society, but its energy consumption is a growing concern. The Internet of Things (IoT) exemplifies this challenge, with billions of devices relying on C for energy efficiency. Outside the IoT domain, Python is becoming increasingly popular despite being one of the least energy-efficient programming languages (PLs). Optimization is therefore essential to reduce energy consumption in both cases: (i) in the IoT, where despite the use of C, the large number of devices leads to high energy consumption, and (ii) in Python, due to its high energy consumption. This paper evaluates 23 optimization techniques in C and Python, taking into account the choice of compiler (O0/O3 in C and Interpreter/Nuitka in Python). The results show that more techniques have a positive impact (reducing energy) in Python than in C, but the energy consumption of Python remains significantly higher, requiring further optimizations to narrow the gap.

Keywords: Software · Energy consumption · Optimization

1 Introduction

The global drive towards sustainability has extended to information technology (IT), leading to the concept of Green IT. Although initial efforts within this domain focused mainly on improving the energy efficiency of hardware systems, the importance of software in achieving environmental goals has become increasingly evident [13]. *Green in software* [1] seeks to minimize the environmental footprint of software throughout its life cycle (as explored in this paper). From this perspective, software energy consumption is significantly impacted by the choices that developers make during the development process [16].

As the Internet of Things (IoT) becomes a transformative force, the need for energy-efficient software becomes even more critical. In this context, C plays a key role due to its low-level control over system resources, making it one of the most energy-efficient languages. However, outside the IoT domain, Python has become one of the most popular programming languages according to the Popularity of Programming Languages (PYPL) index, despite its higher energy consumption [7]. In this paper, based on the need for highly efficient code in IoT environments and the results of [7], we have decided to evaluate the impact of code optimization techniques for both C and its Python equivalent. This large difference in energy consumption between the two languages provides a compelling rationale for our investigation of optimization techniques that could potentially improve the energy efficiency of Python to levels comparable to C. In addition to code optimization techniques, we will explore other optimizations. Specifically, for C, we will examine the impact of optimization levels O0 (no optimization) and O3 (highest optimization) on energy consumption. O1 and O2 have been excluded from this study as they represent intermediate levels of optimization that do not produce significant results. For Python, we chose Nuitka because it applies advanced compile-time optimizations, such as constant folding, constant propagation, built-in call prediction, and conditional statement prediction, which make it more suitable compared to other compilers that do not offer these optimizations. This dual approach aims to answer five research questions:

RQ1) How do efficient coding techniques impact energy consumption in C?
RQ2) How does optimization level impact energy consumption in C?
RQ3) How do efficient coding techniques impact energy consumption in Python?
RQ4) How does the Nuitka compiler impact energy consumption in Python?
RQ5) How do coding techniques impact energy use in C and Python?

To address the stated research questions, 23 software optimization techniques that can be implemented in both languages have been selected from [3–5].

The remainder of the paper is structured as follows: Sect. 2 provides an overview of the related work; Sect. 3 describes the FEETINGS framework, which provides the hardware measurement environment used in the study; Sect. 4 presents the results obtained and the threats to validity are tackled in Sect. 5; and finally, Sect. 6 summarizes the conclusions and future work.

2 Related Work

The comparison of programming languages in terms of energy efficiency has been a topic of significant interest. The study in [7] emphasizes that language choice strongly influences energy consumption. Based on a software-based study that estimated energy use, [7] adopts a hardware-based approach to analyze energy consumption. Compiled languages, such as C and C++, consistently show superior energy efficiency, while interpreted languages like Python, Ruby, and Perl exhibit high energy consumption. Focusing on Haskell, [10] study the impact of compiler optimizations on energy efficiency, finding that enabling -O2 optimizations improved efficiency in 24% of cases.

Specific implementations within a programming language also impact energy efficiency. In C, [8] shows that switching complexity has little effect on energy consumption, while randomness generation significantly increases energy costs. Power measurements were recorded using National Instruments Measurement & Automation Explorer (NI MAX). In line with this, using the widely used Qualcomm Trepn Profiler, [6] shows that refactoring individual energy-intensive code patterns in Android applications improves energy efficiency, while combining refactorings yields mixed results. Similarly, design patterns can influence energy consumption, as shown by [14], which highlights that their general use often leads to a negative impact on energy consumption.

This study builds on previous research examining the impact of software optimization techniques in C, including 26 techniques tested in various environments, such as Raspberry Pi models (IoT device) [3,4], and High Performance Computing (HPC) infrastructures [5]. These techniques were evaluated for their effectiveness in reducing execution time and this study also evaluates the impact of these techniques on energy consumption. For this purpose, we have selected optimization techniques from previous research that can be implemented in both C and Python (code available in the experimental package [15]). Table 1 presents the classification of the selected techniques, which are grouped according to [3–5] as follows: (1) data/memory structure, (2) algorithm implementation, (3) arithmetic/logic operations, (4) loops.

Table 1. Techniques classified by their main characteristics [3–5]

Tech.	Technique name	(1)	(2)	(3)	(4)
T4	Cascaded function calls				x
T7	Common subexpression elimination	x	x		
T8	Mapping structures	x	x		
T9	Dead code elimination		x		
T10	Exception handling		x		x
T11	Global variables within loops	x			x
T12	Function inlining		x		
T13	Global variables	x			x
T16	Division by a power-of-two denominator			x	
T17	Multiplication by a power-of-two factor			x	
T19	Loop countdown		x		x
T20	Loop unrolling		x		x
T25	Linear search		x		x
T26	Invariant if statements within loops		x		x
T28	Loop fission				x
T29	Range check			x	
T31	Buffer IO	x			
T32	Look-up table	x			
T33	Tail function calls	x			
T36	Loop exit using break				x
T40	Loop jamming				x
T44	Lazy evaluation			x	
T49	Sentinels			x	

3 Method: FEETINGS

To ensure the reliability and replicability of our study, we adopted FEETINGS [11], which promotes reliable data collection and analysis of energy consumption. Energy consumption is measured using EET, a hardware-based measurement instrument with a sampling frequency of 100 Hz. It is designed to measure by means of sensors the total energy consumption of the Device Under Test (DUT), which refers to the computer where the software is being executed, as well as of its individual components (processor, graphics card, hard disk, and monitor). The specifications of the DUT used in the study are: Gigabyte H410M S2H V2 motherboard, an i7 10700 2.9 GHz processor, 2 modules of 8 GB DDR4 Crucial 2666 MHz CL19 RAM, ATI Sapphire Radeon X1950 GT graphics card, Western Digital Blue 500GB SSD hard drive, 3Go PS580S 580 W power supply and Ubuntu 20.04.2 LTS operating system.

To carry out the experiments, each test was run within the algorithm loop for at least 100 million iterations to ensure that the runtime of the analyzed code reached a measurable value. Each test case was executed 30 times, providing a sufficient sample size to ensure reliability and minimize outliers, with the final result being the average of these measurements [9]. In the case of C, the code was compiled using GCC version 7.2.0 with two levels of optimization: O0 and O3. For Python, the experiments used Python 3.11.11 with the default interpreter, as well as compiled versions of the code using Nuitka 2.5.6, which uses GCC 10.3.0 internally to generate optimized binaries.

4 Results

This section presents the energy consumption results obtained by the execution of experiments on the DUT. The experimental package can be found in the repository of this study [15].

4.1 RQ1. Impact of Techniques in C

Table 2 presents the results for the standard code alongside those for the efficient code, which was derived by applying the corresponding technique to the standard version. To confirm whether the differences obtained could be considered statistically significant and given the non-normality of the data, the non-parametric Mann-Whitney test was applied and, as a result, the effect size value [2] was calculated, as shown in Table 2 (details can be found in the experimental package [15]).

The results vary depending on the level of optimization. For O0, the technique with the greatest energy reduction is T10, at 99.69%. Of the 23 techniques tested, 15 reduced energy consumption by an average of 28.83%, 2 showed minimal changes (reductions ranging from 0.17% to 0.52%), and 6 increased consumption by an average of 45.69%. In particular, T28 and T31 led to significant increases of 101.56% and 113.02%, respectively (see Table 2). When changing

Table 2. Impact for all techniques in C

Tech.	Optimization level O0				Optimization level O3			
	Standard (J)	Efficient (J)	Impact (%)	Effect size	Standard (J)	Efficient (J)	Impact (%)	Effect size
T4	221.56	142.38	−35.74	large	118.66	21.32	−82.04	large
T7	63.53	27.30	−57.03	large	63.10	28.42	−54.97	large
T8	243.32	288.48	18.56	large	1126.58	226.67	−79.88	large
T9	7.34	6.50	−11.43	medium	6.30	6.11	−3.09	small~
T10	4730.47	14.45	−99.69	large	4750.52	6.15	−99.87	large
T11	70.46	67.04	−4.86	large	24.71	17.92	−27.46	large
T12	23.53	18.70	−20.52	large	7.94	8.54	7.55	small~
T13	418.63	416.47	−0.52	small~	294.70	261.94	−11.12	large
T16	20.51	18.56	−9.49	medium	8.75	9.00	2.78	small~
T17	18.55	19.58	5.57	small~	10.67	9.18	−13.93	small
T19	673.60	887.27	31.72	large	46.14	50.24	8.89	large
T20	436.98	145.37	−66.73	large	26.87	44.19	64.47	large
T25	900.23	729.22	−19.00	large	192.98	145.28	−24.72	large
T26	946.51	944.88	−0.17	small~	238.58	26.22	−89.01	large
T28	344.20	693.76	101.56	large	35.94	33.66	−6.35	medium
T29	14.34	13.90	−3.07	small~	6.45	6.76	4.84	small~
T31	978.72	2084.88	113.02	large	756.50	1718.98	127.23	large
T32	18.96	15.87	−16.29	large	7.14	7.62	6.68	small
T33	24.98	25.90	3.71	small~	20.33	21.55	5.97	small
T36	178.74	171.02	−4.32	large	5.30	5.83	10.07	small
T40	625.09	542.63	−13.19	large	263.63	52.14	−80.22	large
T44	4.06	3.53	−13.14	small	3.66	4.22	15.29	medium
T49	417.29	175.73	−57.89	large	7.25	40.79	462.27	large

Effect sizes with ~ indicate that $p > 0.05$, meaning the null hypothesis cannot be rejected, and thus, no significant differences exist.

to O3, more techniques experienced significant reductions in energy consumption, such as T4 (82.04%), T8 (79.88%), T10 (99.87%), T26 (89.01%), and T40 (80.22%), covering different categories (see Table 1). Despite fewer techniques achieving a reduction compared to O0, the reductions were greater, with an average of 47.72%, exceeding O0. T31 and T49 showed significant increases in energy consumption, with 127.23% and 462.27%, respectively (see Table 2).

Some techniques that reduce energy consumption at the optimization level O0 are also effective in O3, such as T4, T7, T9, T10, T11, and T25, with changes in their impact levels that do not necessarily change from one extreme to another (see Table 2). However, this is not always the case. For example, T49, which has a moderate energy reduction at O0, shows a significant energy increase at O3. This implies that a technique that reduces energy consumption at O0 may not necessarily yield the same benefits at O3, and vice versa.

4.2 RQ2. Impact of the Optimization Level in C

Table 3 shows the results for the optimization levels O0 and O3, where all optimizations are applied. In 89% cases, a reduction in energy consumption is observed for standard and efficient codes, with the reductions primarily moderate to significant.

Table 3. Impact for optimization level O0 and O3 in C

Tech.	Standard				Efficient			
	Opt. O0 (J)	Opt. O3 (J)	Impact (%)	Effect size	Opt. O0 (J)	Opt. O3 (J)	Impact (%)	Effect size
T4	221.56	118.66	−46.44	large	142.38	21.32	−85.03	large
T7	63.53	63.10	−0.67	small∼	27.30	28.42	4.08	medium
T8	243.32	1126.58	363.00	large	288.48	226.67	−21.42	large
T9	7.34	6.30	−14.07	medium	6.50	6.11	−5.99	small∼
T10	4730.47	4750.52	0.42	small∼	14.45	6.15	−57.41	large
T11	70.46	24.71	−64.93	large	67.04	17.92	−73.26	large
T12	23.53	7.94	−66.24	large	18.70	8.54	−54.32	large
T13	418.63	294.70	−29.60	large	416.47	261.94	−37.11	large
T16	20.51	8.75	−57.32	large	18.56	9.00	−51.53	large
T17	18.55	10.67	−42.49	large	19.58	9.18	−53.11	large
T19	673.60	46.14	−93.15	large	887.27	50.24	−94.34	large
T20	436.98	26.87	−93.85	large	145.37	44.19	−69.60	large
T25	900.23	192.98	−78.56	large	729.22	145.28	−80.08	large
T26	946.51	238.58	−74.79	large	944.88	26.22	−97.23	large
T28	344.20	35.94	−89.56	large	693.76	33.66	−95.15	large
T29	14.34	6.45	−55.07	large	13.90	6.76	−51.40	large
T31	978.72	756.50	−22.71	large	2084.88	1718.98	−17.55	large
T32	18.96	7.14	−62.33	large	15.87	7.62	−52.00	large
T33	24.98	20.33	−18.59	large	25.90	21.55	−16.81	large
T36	178.74	5.30	−97.04	large	171.02	5.83	−96.59	large
T40	625.09	263.63	−57.83	large	542.63	52.14	−90.39	large
T44	4.06	3.66	−9.79	small∼	3.53	4.22	19.73	medium
T49	417.29	7.25	−98.26	large	175.73	40.79	−76.79	large

Effect sizes with ∼ indicate that $p > 0.05$, meaning the null hypothesis cannot be rejected, and thus, no significant differences exist.

For the standard code, only T8 shows a significant energy increase of 362.55%. On the other hand, for the efficient code, T7 and T44 showed a slight increase in the energy consumption of 4.08% and 19.73%, respectively. Techniques that increase energy consumption behave differently between the standard and efficient code, as the optimization level depends on the code itself. T8 is an example of this behavior: the standard code shows a significant increase in energy consumption (363.00%) at O3, while the efficient code reduces its energy consumption by 21.42%.

4.3 RQ3. Impact of Techniques in Python

Table 4 shows the results for the standard code alongside those for the efficient code, in a similar way to what was previously done for C, but with the results obtained for the Python interpreter and Nuitka. Here, the results show significantly higher energy consumption compared to C, as Python generally consumes more energy than C because of its higher-level nature, which introduces more overhead and reduces efficiency in resource management.

When using the Python interpreter to run the Python code, techniques T4, T7, T8, T10, T11, T26, T40, and T49 show a moderate energy reduction, with an average reduction of 28.13%. Of 23 techniques, 16 experienced a reduction

Table 4. Impact for all techniques in Python

Tech.	Interpreter				Nuitka			
	Standard (J)	Efficient (J)	Impact (%)	Effect size	Standard (J)	Efficient (J)	Impact (%)	Effect size
T4	5830.42	2901.13	−50.24	large	8100.00	2634.99	−67.47	large
T7	1405.79	955.12	−32.06	large	1261.44	722.59	−42.72	large
T8	1393.72	408.05	−70.72	large	1027.67	477.43	−53.54	large
T9	348.37	265.03	−23.92	large	157.98	127.74	−19.14	large
T10	1596.32	726.44	−54.49	large	841.16	262.18	−68.83	large
T11	2697.65	1833.46	−32.03	large	1618.23	1733.50	7.12	large
T12	549.22	441.43	−19.63	large	290.41	249.06	−14.24	large
T13	14345.32	13396.95	−6.61	large	13134.52	13182.98	0.37	large
T16	494.35	434.30	−12.15	large	254.60	240.77	−5.43	large
T17	439.66	467.30	6.29	large	236.39	289.03	22.27	large
T19	8087.84	8009.26	−0.97	small∼	7902.13	7644.86	−3.26	large
T20	4211.31	3744.15	−11.09	large	3894.97	4441.61	14.03	large
T25	10830.99	9762.90	−9.86	large	9479.08	8230.04	−13.18	large
T26	11806.91	8477.69	−28.20	large	10351.72	8654.01	−16.40	large
T28	12319.90	15106.71	22.62	large	12138.42	14212.57	17.09	large
T29	317.72	321.51	1.19	small∼	173.69	216.47	24.64	large
T31	10367.75	21580.19	108.15	large	27458.03	29191.64	6.31	medium
T32	559.13	498.30	−10.88	large	242.13	303.89	25.51	large
T33	609.18	598.93	−1.68	small	339.84	327.38	−3.66	medium
T36	4559.29	3859.49	−15.35	large	3004.91	2940.67	−2.14	medium
T40	12280.55	7948.52	−35.28	large	9262.39	9118.77	−1.55	medium
T44	351.30	366.02	4.19	medium	143.54	140.52	−2.11	small∼
T49	5814.26	3628.37	−37.60	large	5248.71	3528.80	−32.77	large

Effect sizes with ∼ indicate that $p > 0.05$, meaning the null hypothesis cannot be rejected, and thus, no significant differences exist.

in energy consumption, ranging from 6.61% to 70.72%, consisting mostly of slight reductions. Only three techniques showed a slight increase, with an average increase of 11.03%, and one technique, T31, showed a significant increase of 108.15%, similar to C. For the Nuitka compiler, moderate energy savings are observed in some cases, similar to when using the Python interpreter, except for T11, T26, and T40. In particular, T11 shows a slight increase in energy consumption of 7.12%, T6 reduces energy consumption by 16.40%, but to a lesser extent compared to the use of the Python interpreter, and T40 does not show notable differences, with a reduction of 1.55%. Fewer techniques result in energy savings, with 12 techniques showing a reduction compared to 16 when using the Python interpreter. However, the average reduction is 28.39%, slightly higher than the one observed with Nuitka. In addition, more techniques lead to higher energy consumption, but the average increase is lower (16.71%) than that of the Python interpreter (35.31%), mainly due to the increase caused by T31 in the Python interpreter.

4.4 RQ4. Impact of Nuitka Compiler in Python

Table 5 shows the results for the Python interpreter and Nuitka. In 76% of the cases, a reduction in energy consumption is observed, mainly slight and mod-

Table 5. Impact for Interpreter and Nuitka in Python

Tech.	Standard				Efficient			
	Interpreter (J)	Nuitka (J)	Impact (%)	Effect size	Interpreter (J)	Nuitka (J)	Impact (%)	Effect size
T4	5830.42	8100.00	38.93	large	2901.13	2634.99	−9.17	large
T7	1405.79	1261.44	−10.27	large	955.12	722.59	−24.35	large
T8	1393.72	1027.67	−26.26	large	408.05	477.43	17.00	large
T9	348.37	157.98	−54.65	large	265.03	127.74	−51.80	large
T10	1596.32	841.16	−47.31	large	726.44	262.18	−63.91	large
T11	2697.65	1618.23	−40.01	large	1833.46	1733.50	−5.45	large
T12	549.22	290.41	−47.12	large	441.43	249.06	−43.58	large
T13	14345.32	13134.52	−8.44	large	13396.95	13182.98	−1.60	large
T16	494.35	254.60	−48.50	large	434.30	240.77	−44.56	large
T17	439.66	236.39	−46.23	large	467.30	289.03	−38.15	large
T19	8087.84	7902.13	−2.30	medium	8009.26	7644.86	−4.55	large
T20	4211.31	3894.97	−7.51	large	3744.15	4441.61	18.63	large
T25	10830.99	9479.08	−12.48	large	9762.90	8230.04	−15.70	large
T26	11806.91	10351.72	−12.32	large	8477.69	8654.01	2.08	medium
T28	12319.90	12138.42	−1.47	medium	15106.71	14212.57	−5.92	large
T29	317.72	173.69	−45.33	large	321.51	216.47	−32.67	large
T31	10367.75	27458.03	164.84	large	21580.19	29191.64	35.27	large
T32	559.13	242.13	−56.69	large	498.30	303.89	−39.01	large
T33	609.18	339.84	−44.21	large	598.93	327.38	−45.34	large
T36	4559.29	3004.91	−34.09	large	3859.49	2940.67	−23.81	large
T40	12280.55	9262.39	−24.58	large	7948.52	9118.77	14.72	large
T44	351.30	143.54	−59.14	large	366.02	140.52	−61.61	large
T49	5814.26	5248.71	−9.73	large	3628.37	3528.80	−2.74	large

Effect sizes with ∼ indicate that $p > 0.05$, meaning the null hypothesis cannot be rejected, and thus, no significant differences exist.

erate. For the standard code, there are few cases where an increase in energy consumption occurs, such as T4, with 38.93%, and T31, with 164.84%. In the case of the efficient code, more techniques show an increase in energy consumption compared to the standard code (4 versus 2), although these increases are mainly slight. As a result, the average increase in energy is 21.41%, compared to 101.88% in the standard code. Techniques that increase energy consumption in the efficient code are similar to those in the standard code, except for T4, along with new ones such as T8, T20, and T40.

4.5 RQ5. Comparison of the Impact of Techniques in C and Python

There are slightly more techniques with energy reductions in Python than in C (61% compared to 59%) and significantly fewer with increases (24% compared to 35%). This suggests that the selected set of techniques has a better energy impact in Python than in C. However, Python's overall energy consumption remains much higher than C (see Tables 2 and 4), suggesting that further optimization is needed to close the gap in terms of energy consumption. To determine if the observed differences were statistically significant, and considering the non-normality of the data, the non-parametric Mann-Whitney test was used to compare C and Python. Since the resulting p-value was less than 0.05, the null hypothesis was rejected, indicating that there are significant differences (for more details, see the experimental package [15]).

5 Threats to Validity

To minimize threats to construct validity, we used the EET hardware device, which offers more accurate energy consumption results than software tools. These results are specific to this device and may vary with others, although significant differences are unlikely, since a previous study validated EET against a gold standard [12]. To ensure internal validity, the DUT was restarted and set to the same state before each test, background processes were minimized, and a proportional stabilization time was allowed before each test. For external validity, a representative set of optimization techniques from different categories was selected (see Table 1). However, additional techniques exist, so the results may vary if other optimizations are considered.

6 Conclusions

This study highlights the importance of optimization techniques in reducing energy consumption. Both C and Python benefit from these techniques, with reductions of up to 99.87% in C and 70.72% in Python. Developers should prioritize energy efficiency in their coding practices, not only in Python, known for its inefficiency, but also in C, despite being more efficient. Compiler optimizations, such as those in C and Nuitka for Python, also play a key role in improving energy efficiency. A multi-faceted approach, combining efficient coding and compiler optimizations, is essential for more sustainable software.

Future work should explore more programming languages and apply both existing and language-specific optimization techniques, automate code refactoring for energy-efficient software, and test these techniques on real-world projects.

Acknowledgements. This work has been supported by the following projects: OASSIS (PID2021-122554OB-C31/ AEI/10.13039/ 501100011033/FEDER, UE); EMMA (Project SBPLY/21/180501/000115, funded by CECD (JCCM) and FEDER funds); PLAGEMIS (TED2021-129245B-C22 funded by MCIN/AEI/ 10.13039/501100011033 and European Union NextGenerationEU/PRTR); Financial support for the execution of applied research projects, within the framework of the UCLM Own Research Plan, co-financed at 85% by the European Regional Development Fund (FEDER) UNION (2022-GRIN-34110).

References

1. Calero, C., Piattini, M. (eds.): Green in Software Engineering. Springer, Cham (2015). https://doi.org/10.1007/978-3-319-08581-4
2. Cohen, J.: Statistical Power Analysis for the Behavioral Sciences, 2nd edn. Routledge, New York. (2013). https://doi.org/10.4324/9780203771587
3. Corral-García, J., González-Sánchez, J.L., Pérez-Toledano, M.Á.: Evaluation of strategies for the development of efficient code for raspberry pi devices. Sensors **18**(11), 4066 (2018). https://doi.org/10.3390/s18114066. https://www.mdpi.com/1424-8220/18/11/4066
4. Corral-García, J., Lemus-Prieto, F., González-Sánchez, J.L., Pérez-Toledano, M.Á.: Analysis of energy consumption and optimization techniques for writing energy-efficient code. Electronics **8**(10), 1192 (2019). https://doi.org/10.3390/electronics8101192. https://www.mdpi.com/2079-9292/8/10/1192
5. Corral-García, J., Lemus-Prieto, F., Pérez-Toledano, M.Á.: Efficient code development for improving execution performance in high-performance computing centers. J. Supercomput. **77**(4), 3261–3288 (2021). https://doi.org/10.1007/s11227-020-03382-z
6. Couto, M., Saraiva, J., Fernandes, J.P.: Energy refactorings for android in the large and in the wild. In: 2020 IEEE 27th International Conference on Software Analysis, Evolution and Reengineering (SANER), pp. 217–228 (2020). https://doi.org/10.1109/SANER48275.2020.9054858. https://ieeexplore.ieee.org/document/9054858
7. Gordillo, A., et al.: Programming languages ranking based on energy measurements. Softw. Qual. J. **32**(4), 1539–1580 (2024). https://doi.org/10.1007/s11219-024-09690-4
8. Jain, R., Molnar, D., Ramzan, Z.: Towards understanding algorithmic factors affecting energy consumption: switching complexity, randomness, and preliminary experiments. In: Proceedings of the 2005 Joint Workshop on Foundations of Mobile Computing, DIALM-POMC '05, pp. 70–79. Association for Computing Machinery, New York (2005). https://doi.org/10.1145/1080810.1080823
9. Kern, E., et al.: Sustainable software products—Towards assessment criteria for resource and energy efficiency. Fut. Gener. Comput. Syst. **86**, 199–210 (2018). https://doi.org/10.1016/j.future.2018.02.044. https://www.sciencedirect.com/science/article/pii/S0167739X17314188
10. Kirkeby, M.H., Santos, B., Fernandes, J.P., Pardo, A.: Compiling haskell for energy efficiency: empirical analysis of individual transformations. In: Proceedings of the 39th ACM/SIGAPP Symposium on Applied Computing, SAC '24, pp. 1104–1113. Association for Computing Machinery, New York (2024). https://doi.org/10.1145/3605098.3635915
11. Mancebo, J., Calero, C., Garcia, F., Moraga, M.A., Garcia-Rodriguez de Guzman, I.: FEETINGS: framework for energy efficiency testing to improve environmental goal of the software. Sustain. Comput. Inf. Syst. **30**, 100558 (2021). https://doi.org/10.1016/j.suscom.2021.100558. https://www.sciencedirect.com/science/article/pii/S2210537921000494
12. Mancebo, J., et al.: Assessing the sustainability of software products—a method comparison. In: Schaldach, R., Simon, K.-H., Weismüller, J., Wohlgemuth, V. (eds.) Advances and New Trends in Environmental Informatics. PI, pp. 1–15. Springer, Cham (2020). https://doi.org/10.1007/978-3-030-30862-9_1
13. Pinto, G., Castor, F.: Energy efficiency: a new concern for application software developers. Commun. ACM **60**(12), 68–75 (2017). https://doi.org/10.1145/3154384

14. Poy, O., Moraga, M.Á., García, F., Calero, C.: Impact on energy consumption of design patterns, code smells and refactoring techniques: a systematic mapping study. J. Syst. Softw. **222**, 112303 (2025). https://doi.org/10.1016/j.jss.2024.112303. https://www.sciencedirect.com/science/article/pii/S0164121224003479
15. Pulido, C., García, F.O., Ángeles Moraga, M., Calero, C.: GrupoAlarcos/Evaluating-Programming-Optimization-Techniques-in-C-and-Python–Impact-on-Energy-Consumption (2025). https://github.com/GrupoAlarcos/Evaluating-Programming-Optimization-Techniques-in-C-and-Python--Impact-on-Energy-Consumption
16. Zhang, C., Hindle, A., German, D.M.: The impact of user choice on energy consumption. IEEE Softw. **31**(3), 69–75 (2014). https://doi.org/10.1109/MS.2014.27. https://ieeexplore.ieee.org/document/6756706

An Analysis of Resilience in Digital Business Ecosystems

Beāte Krauze

Institute of Information Technology, Riga Technical University, 6A Kipsalas Street, Riga 1048, Latvia
`beate.krauze@edu.rtu.lv`

Abstract. The resilience and compliance of Digital Business Ecosystems (DBEs) are critical key priorities, particularly as digitalisation expands and organisations face increasing regulatory scrutiny. The resilience of these ecosystems – their ability to adapt, recover, and continue functioning during disruptions – is an essential focus. The European Union's regulations and directives emphasise the necessity of strategies to ensure DBE resilience. These legislative enactments demand compliance to strict requirements for operational resilience, cybersecurity practices, and incident reporting, reflecting the growing complexity of regulatory expectations. This paper aims to review the literature on resilience-related challenges. This study provides valuable insights into future research directions of DBEs.

Keywords: Digital Business Ecosystem · Technology · Resilience · EU Law · Cybersecurity

1 Introduction

While researchers have performed systematic literature reviews to explore Digital Business Ecosystems (DBEs), a gap remains in synthesising and evaluating experiences related to sustaining resilience in these ecosystems [1–4]. Governance, regulations, security of DBE and digital infrastructure mobilization were identified as some of the most important aspects and less-researched areas [1]. This mapping study focuses on extant studies on these topics, namely: resilience and resilience requirements from the legislative enactments, emphasising the role of digital infrastructure mobilisation during disruptions.

The main research questions motivating this study are:

RQ1: What are the key challenges associated with maintaining resilience in DBE?
RQ2: How do resilience challenges relate to the European regulatory environment?

To address these research questions, this paper is organized as follows. An overview of the DBE concept is first presented. This is followed by an explanation of the research methodology, including the literature search, article selection, refinement, and analysis approaches. The findings are then presented to address the research questions. Subsequently, DBE resilience challenges are highlighted. The paper concludes with key conclusions and future research directions.

2 Background

DBE is a dynamic, interconnected, and technology-driven environment, comprising individuals, organisations and digital entities that co-create value using information and communication technologies [1, 5]. The concept has evolved from traditional business ecosystems by integrating digital technologies, automation, and real-time data exchange, fostering more interconnected and interdependent networks of digital and physical entities.

Resilience in DBEs refers to the ability of an ecosystem to withstand, adapt to, and recover from disruptions while maintaining continuous operations and minimising negative impacts [6]. Resilience is achieved through a combination of technological, organisational, and regulatory measures, including cybersecurity measures, risk management, and compliance with legal requirements.

As these ecosystems grow in scope and complexity, the need for resilience becomes crucial, particularly in the context of security, regulations and operations [7, 8]. The EU has implemented several regulations to enhance the resilience, focusing on cybersecurity and operational stability.

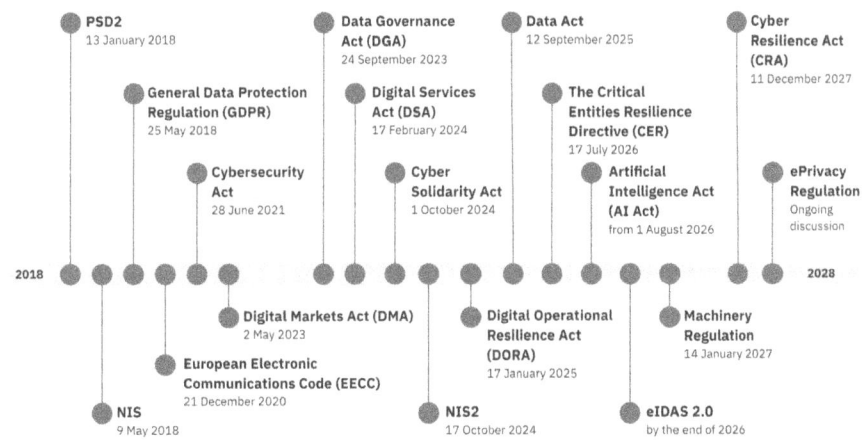

Fig. 1. Dispersion of legislative enactments 2018 – 2028 per application date

The timeline (Fig. 1) highlights a dense and extensive sequence of European regulations and directives focused on digital services, cybersecurity, data protection, and emerging technologies. While it demonstrates the EU's commitment to addressing the complexities of the digital age, the volume and pace of implementation may feel overwhelming for organisations, policymakers, and stakeholders.

3 Research Methodology

To address the RQs comprehensively, an initial literature pool is constructed by examining four databases of scientific literature. The following criteria are applied to select the initial pool of studies (see Fig. 2):

- Databases: Scopus, Science Direct, Web of Science and ACM Digital Library;
- Range: 2015 - 2025 as the year of publication;
- Language: English;
- Subject area: Computer science;
- Paper type: Article, Conference paper;
- Search string content applied for title, abstract, or keywords: ((("digital business ecosystem" OR "DBE") AND resilience) OR ((("digital business ecosystem" OR "DBE") AND cyber) OR ((("digital business ecosystem" OR "DBE") AND regulation OR law OR directive)

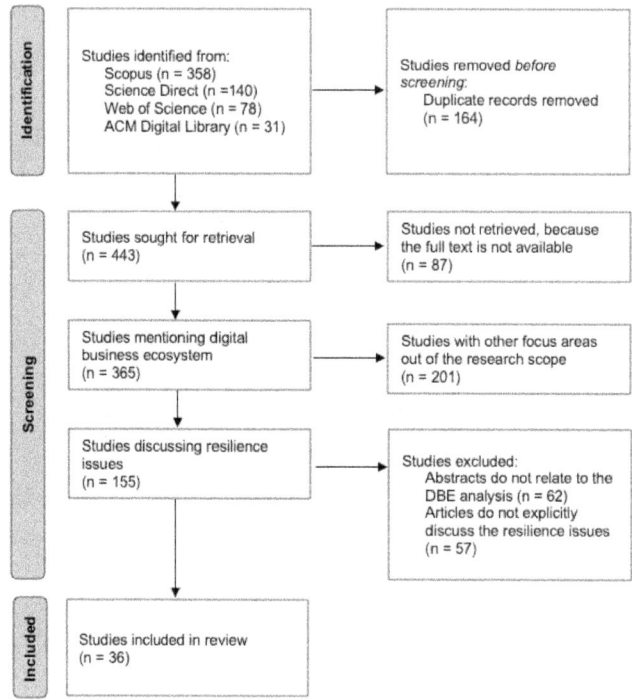

Fig. 2. PRISMA 2020 flow diagram for the literature search process.

The trend of increasing number of articles each year (Fig. 3) reflects the growing emphasis on incorporating recent and relevant papers. This shift indicates a focus on addressing current challenges.

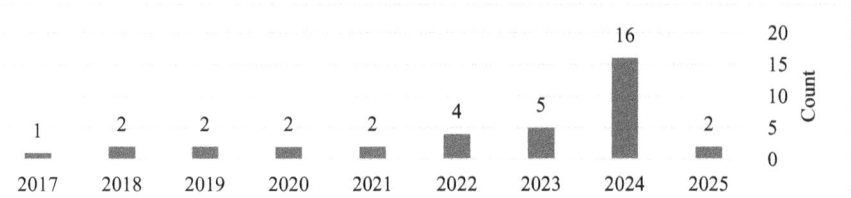

Fig. 3. Dispersion of papers published per

To minimise threats to validity, the following strategies were applied: (1) the selection of studies followed a systematic process using predefined search strings and inclusion/exclusion criteria to ensure relevant research contributions, (2) the study encompasses a wide range of sources, including journal articles and conference proceedings, ensuring a broad representation of perspectives on DBE resilience, (3) a well-defined methodology was applied, including the use of established databases and structured analysis techniques to obtain the results and given classification of the resilience challenges.

4 Results

This subsection addresses the research question **RQ1**: "What are the key challenges associated with maintaining resilience in DBE?".

The classification of challenges is supported by a structured review process. The papers were systematically processed using predefined selection criteria to obtain this structure. The analysis included identifying key resilience challenges and mapping them to relevant legislative enactments, ensuring a comprehensive understanding of how resilience is impacted within DBEs. The categorisation of challenges was derived through thematic coding and synthesis of recurring themes from the literature. This included analysing previous studies, extracting key topics, and grouping them based on their conceptual similarities and practical implications in DBE resilience.

As presented in Fig. 4, DBE resilience challenges can be classified into the following categories.

As summarised in Table 1, while some challenges like technological integration, supply chain and operational resilience, cybersecurity and digital trust challenges and regulatory, compliance challenges are extensively studied, they have relatively few solutions, potentially indicating gaps in research or practical application.

Further in this section resilience challenges are explored deeper into the seven key resilience challenges identified above, providing detailed insights into their nature and implications.

Technological Integration and Interoperability Challenges. The challenges include managing the shift from traditional linear industry structures to layered modular architectures, which require navigating technical and strategic control points, such as modularity, scalability, and customer access [16]. Furthermore, there are challenges in the context of digital transformation, i.e. difficulties in integrating digital technologies due to a lack of

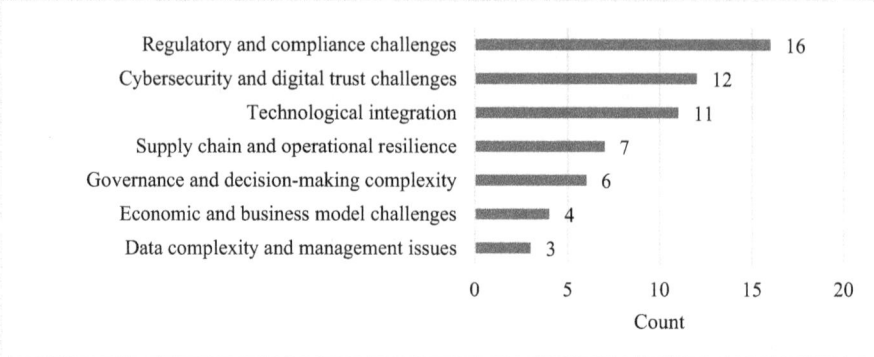

Fig. 4. Number of studies for each of the challenges

Table 1. Resilience challenges.

	Resilience challenge	Mentioned and analysed	Solution technique provided
RC1	Technological integration and interoperability challenges	[2, 3, 7, 9–16]	[13, 15, 16]
RC2	Supply chain and operational resilience	[6, 10, 17, 18]	[6, 10]
RC3	Cybersecurity and digital trust challenges	[9, 12, 13, 19–27]	[9, 13, 19, 21, 22]
RC4	Data complexity and management issues	[10, 15, 21]	[15, 21]
RC5	Regulatory and compliance challenges	[5, 19, 20, 23–33]	[19, 29]
RC6	Governance and decision-making complexity	[3, 15, 21, 30, 34, 35]	[15, 21, 30, 34, 35]
RC7	Economic and business model challenges	[7, 12, 18, 29]	[18, 29]

expertise, resistance to change, and inadequate infrastructure [11]. Challenges also arise from poor ecosystem design, such as inadequate configuration, governance choices, and monetisation strategies, which often lead to an imbalance between openness and control [2].

Supply Chain and Operational Resilience. Critical infrastructures and essential services exemplify distinct cases of DBE with specific requirements during emergency, crisis, and disasters. These include the increasing scale and unpredictability of natural and human-made events, such as pandemics, cyberattacks, and geopolitical conflicts, which complicate prevention and management. Gaps exist in early warning systems, real-time monitoring, and multidisciplinary collaboration, limiting proactive and effective disaster resilience across diverse domains [10, 36].

Cybersecurity and Digital Trust Challenges. Resilience challenges include the increasing complexity and frequency of cyber threats, which necessitate continuous adaptation and response strategies. Organisations face difficulties in integrating resilience into their existing cybersecurity frameworks due to a lack of standardised definitions and methodologies [9, 12, 22]. Challenges in digital security by design, highlights the tension between user convenience and security protocols, especially in IoT and cloud computing. It underscores the need for comprehensive integration of advanced security measures, such as zero-trust architectures and blockchain, to address vulnerabilities in DBE [22].

Data Complexity and Management Issues. Data complexities, including heterogeneity, decentralisation, and uncertainty, hinder resilience efforts [10]. High-frequency data exchanges can increase the damage caused by tampered or compromised data, affecting operational decisions across the DBE. [21] identifies trust management challenges in cybersecurity, such as difficulties in integrating diverse trust mechanisms, maintaining transparency and accountability, and addressing complex socio-technical interactions.

Regulatory and Compliance Challenges. Legal enactments include frequent updates, regulatory fragmentation, and the introduction of complex and industry-specific requirements. Businesses face significant uncertainty due to constantly changing regulations, making long-term investment planning difficult [20, 24, 25]. The costs associated with compliance, operational disruptions, and adapting to new frameworks impose additional burdens, especially on smaller organisations. These challenges are further escalated by the need for cybersecurity measures to counteract increasingly sophisticated cyber threats [26, 28, 31, 33]. For example, sectors such as open banking and smart cities face difficulties in balancing innovation, security, and privacy requirements, while maritime transport and medical data exchange struggle with outdated security measures and complex regulatory compliance [23]. Lastly, unintended consequences of state interventions lead to major challenge, i.e. the knowledge gap of regulators, who lack the dynamic information needed to adapt policies [27].

Governance and Decision-Making Complexity. Decision-making in DBE is often distributed among various actors (service providers, platform owners, regulators). This can obscure accountability for resilience measures. The conducted survey reveals that current research on DBEs recognizes the fundamental roles in service and product exchange [37]. Additionally, it identifies several other roles specific to DBE initialization, actor onboarding, platform maintenance, knowledge dissemination, policy and ethics management, quality assurance of exchanged resources, and more.

Economic and Business Model Challenges. Key challenges in include managing the complexity of federated ecosystems, aligning digital transformation efforts with overarching government strategies, and overcoming inefficiencies such as redundancies and a lack of interoperability across government agencies [7, 18, 29].

5 Discussion

On top of these challenges mentioned above, legislative enactments impose strict obligations related to resilience. Further this subsection presents findings and addresses research question **RQ2:** "How do resilience challenges relate to the European regulatory environment?".

Even though most of the legislative enactments are applied to single organisations, DBEs consist of multiple interconnected entities that depend on shared digital infrastructures, services, and governance structures. This interconnectedness amplifies resilience challenges, as regulatory compliance must be ensured not only at the individual organisational level, but also across the entire ecosystem. The complexity of ensuring synchronised adherence to evolving regulatory landscapes highlights the necessity for collaborative resilience strategies in DBEs. The findings are presented in Table 2.

Table 2. Resilience requirements.

Legislative enactment	Resilience challenge addressed
Cybersecurity Act	RC3
Digital Services Act (DSA)	RC6
Data Governance Act (DGA)	RC4
Cyber Solidarity Act	RC2
NIS2 Directive	RC3
Digital Operational Resilience Act (DORA)	RC1, RC5
Data Act	RC4, RC7
The Critical Entities Resilience Directive (CER)	RC2, RC6
Artificial Intelligence Act (AI Act)	RC1, RC5
Cyber Resilience Act (CRA)	RC3, RC5

6 Conclusion and Further Research

Resilience in DBEs is multidimensional, influenced by complex interdependencies among diverse stakeholders, technological change, and cyber threats. Balancing resilience measures requires coordinated governance, comprehensive risk assessments, and continuous testing practices across the DBE. Further research is required to include: (1) methodologies, tools and practices for resilience testing (e.g., threat modelling) to improve consistency and transparency of assessments specifically across diverse DBEs and (2) metrics such as key performance indicators (KPI) and benchmarks for measuring resilience over time, including both the technical and organisational aspects of DBE preparedness.

This study categorises resilience challenges into focus areas. While some challenges like technological integration, supply chain and operational resilience, cybersecurity and digital trust challenges and regulatory, compliance challenges are extensively studied, they have relatively few solutions, potentially indicating gaps in research or practical application. Further research would benefit from investigating strategies to consolidate multiple regulatory requirements (e.g. DORA, CRA, NIS2) into a cohesive resilience governance framework, minimising duplication and streamlining cross-border compliance.

With the application of DORA (for operational resilience in financial services), the CRA (ensuring the cybersecurity of connected products and software throughout their lifecycle, promoting resilience by design), and NIS2 (improving the cybersecurity and resilience of critical and essential entities across various sectors), organisations may find themselves studying new requirements regularly. Harmonising controls, audits, and reporting obligations can become resource-intensive and requires dedicated compliance structures. By 2030 we can expect new initiatives for collective incident response and resilience at the EU level, potentially tied to shared cybersecurity infrastructure. Legislative proposals such as the AI Act and Data Act may be finalised before 2030, introducing additional cybersecurity and data governance requirements.

References

1. Senyo, P.K., Liu, K., Effah, J.: Digital business ecosystem: literature review and a framework for future research. Int. J. Inf. Manag. **47**, 52–64 (2019). https://doi.org/10.1016/j.ijinfomgt.2019.01.002
2. Coskun-Setirek, A., Carmela Annosi, M., Hurst, W., Dolfsma, W., Tekinerdogan, B.: Architecture and governance of digital business ecosystems: a systematic literature review. Inf. Syst. Manag. **41**, 58–90 (2024). https://doi.org/10.1080/10580530.2023.2194063
3. Tsai, C.H., Zdravkovic, J., Stirna, J.: Modeling digital business ecosystems: a systematic literature review. Complex Syst. Inf. Model. Q. **2022**, 1–30 (2022). https://doi.org/10.7250/csimq.2022-30.01
4. Tan, C., Dhakal, S., Ghale, B.: Conceptualising capabilities and value co-creation in a digital business ecosystem (DBE): a systematic literature review. J. Inf. Syst. Eng. Manag. **5** (2020). https://doi.org/10.29333/jisem/7826
5. Korpela, K., Kuusiholma, U., Taipale, O., Hallikas, J.: A framework for exploring digital business ecosystems. In: Proceedings of the Annual Hawaii International Conference on System Sciences, pp. 3838–3847 (2013). https://doi.org/10.1109/HICSS.2013.37
6. Farmakis, T., Koukopoulos, A., Zois, G., Mourtos, I., Lounis, S., Kalaboukas, K.: Developing a circular and resilient information system: a design science approach. In: International Federation for Information Processing, pp. 64–79 (2024). https://doi.org/10.1007/978-3-031-71622-5_5
7. Gill, A., Hansnata, M.: Digital government ecosystem: adaptive architecture for digital and ICT investment decision making. In: ACM International Conference Proceeding Series, pp. 555–564. Association for Computing Machinery (2024). https://doi.org/10.1145/3657054.3657119
8. Krauze, B., Grabis, J.: A conceptual model of digital immune system to increase the resilience of technology ecosystems. In: Lecture Notes in Business Information Processing, pp. 82–96. Springer Science and Business Media Deutschland GmbH (2024) https://doi.org/10.1007/978-3-031-59465-6_6
9. Tornjanski, V., Glišić, V.: Towards secured digital business ecosystems: from threats to opportunities. In: International Conference on E-Business Technologies (2021)

10. Cao, L.: AI and data science for smart emergency, crisis and disaster resilience. Int. J. Data Sci. Anal. **15**, 231–246 (2023). https://doi.org/10.1007/s41060-023-00393-w
11. Awad, J., Martín-Rojas, R.: Digital transformation influence on organisational resilience through organisational learning and innovation. J. Innov. Entrep. **13** (2024). https://doi.org/10.1186/s13731-024-00405-4
12. Tzavara, V., Vassiliadis, S.: Tracing the evolution of cyber resilience: a historical and conceptual review. Int. J. Inf. Secur. **23**, 1695–1719 (2024). https://doi.org/10.1007/s10207-023-00811-x
13. Hasani, T., O'Reilly, N., Dehghantanha, A., Rezania, D., Levallet, N.: Evaluating the adoption of cybersecurity and its influence on organizational performance. SN Bus. Econ. **3** (2023). https://doi.org/10.1007/s43546-023-00477-6
14. Chen, A., Lin, Y., Mariani, M., Shou, Y., Zhang, Y.: Entrepreneurial growth in digital business ecosystems: an integrated framework blending the knowledge-based view of the firm and business ecosystems. J. Technol. Transfer **48**, 1628–1653 (2023). https://doi.org/10.1007/s10961-023-10027-9
15. Lenkenhoff, K., Wilkens, U., Zheng, M., Süße, T., Kuhlenkötter, B., Ming, X.: Key challenges of digital business ecosystem development and how to cope with them. Procedia CIRP **73**, 167–172 (2018). https://doi.org/10.1016/j.procir.2018.04.082
16. Bohnsack, R., Rennings, M., Block, C., Bröring, S.: Profiting from innovation when digital business ecosystems emerge: a control point perspective. Res. Policy **53** (2024). https://doi.org/10.1016/j.respol.2024.104961
17. Bachtiar, N.K., Setiawan, A., Prastyan, G.A., Kijkasiwat, P.: Business resilience and growth strategy transformation post crisis. J. Innov. Entrep. **12** (2023). https://doi.org/10.1186/s13731-023-00345-5
18. Fornasiero, R., Tolio, T.A.M.: Digital supply chains for ecosystem resilience: a framework for the Italian case. Oper. Manag. Res. (2024). https://doi.org/10.1007/s12063-024-00511-2
19. Proudfoot, J.G., Cram, W.A., Madnick, S.: Weathering the storm: examining how organisations navigate the sea of cybersecurity regulations. Eur. J. Inf. Syst. (2024). https://doi.org/10.1080/0960085X.2024.2345867
20. Bygrave, L.A.: The emergence of EU cybersecurity law: a tale of lemons, angst, turf, surf and grey boxes. Comput. Law Secur. Rev. **56** (2025). https://doi.org/10.1016/j.clsr.2024.106071
21. Pigola, A., de Souza Meirelles, F.: Unraveling trust management in cybersecurity: insights from a systematic literature review. Inf. Technol. Manag. (2024). https://doi.org/10.1007/s10799-024-00438-x
22. Radanliev, P.: Digital security by design. Secur. J. (2024). https://doi.org/10.1057/s41284-024-00435-3
23. Fischer-Hübner, S., et al.: Stakeholder perspectives and requirements on cybersecurity in Europe. J. Inf. Secur. Appl. **61**, 102916 (2021). https://doi.org/10.1016/j.jisa.2021.102916
24. Chiara, P.G.: The cyber resilience act: the EU Commission's proposal for a horizontal regulation on cybersecurity for products with digital elements. Int. Cybersecur. Law Rev. **3**, 255–272 (2022). https://doi.org/10.1365/s43439-022-00067-6
25. Salvaggio, S.A., González, N.: The European framework for cybersecurity: strong assets, intricate history. Int. Cybersecur. Law Rev. **4**, 137–146 (2023). https://doi.org/10.1365/s43439-022-00072-9
26. Markopoulou, D., Papakonstantinou, V., de Hert, P.: The new EU cybersecurity framework: the NIS Directive, ENISA's role and the General Data Protection Regulation. Comput. Law Secur. Rev. **35** (2019). https://doi.org/10.1016/j.clsr.2019.06.007
27. Hodgins, M.W.: The perils of cybersecurity regulation. Rev. Aust. Econ. (2024). https://doi.org/10.1007/s11138-024-00660-4

28. Kianpour, M., Raza, S.: More than malware: unmasking the hidden risk of cybersecurity regulations. Int. Cybersecur. Law Rev. **5**, 169–212 (2024). https://doi.org/10.1365/s43439-024-00111-7
29. Degen, K., Teubner, T.: Wallet wars or digital public infrastructure? Orchestrating a digital identity data ecosystem from a government perspective. Electron. Mark. **34** (2024). https://doi.org/10.1007/s12525-024-00731-1
30. Aldea, A., Kusumaningrum, M.C., Iacob, M.E., Daneva, M.: Modeling and analyzing digital business ecosystems: an approach and evaluation. In: IEEE International Conference on Business Informatics, pp. 156–163. Institute of Electrical and Electronics Engineers Inc. (2018). https://doi.org/10.1109/CBI.2018.10064
31. Buckley, G., Caulfield, T., Becker, I.: "It may be a pain in the backside but..." insights into the resilience of business after GDPR. In: ACM International Conference Proceeding Series, pp. 21–34. Association for Computing Machinery (2022). https://doi.org/10.1145/3584318.3584320
32. Eke, D., Stahl, B.: Ethics in the governance of data and digital technology: an analysis of european data regulations and policies. Dig. Soc. **3** (2024). https://doi.org/10.1007/s44206-024-00101-6
33. Schmittner, C., Veledar, O., Faschang, T., Macher, G., Brenner, E.: Fostering cyber resilience in Europe: an in-depth exploration of the cyber resilience act. In: Yilmaz, M., Clarke, P., Riel, A., Messnarz, R., Greiner, C., Peisl, T. (eds.) European Conference on Software Process Improvement, pp. 390–404. Springer, Cham (2024). https://doi.org/10.1007/978-3-031-71139-8
34. Senyo, P.K., Liu, K., Effah, J.: Towards a methodology for modelling interdependencies between partners in digital business ecosystems. In: IEEE International Conference on Logistics, Informatics and Service Sciences (2017)
35. Grabis, J., Tsai, C.H., Zdravkovic, J., Stirna, J.: Endurant ecosystems: model-based assessment of resilience of digital business ecosystems. In: Lecture Notes in Business Information Processing, pp. 53–68. Springer Science and Business Media Deutschland GmbH (2022). https://doi.org/10.1007/978-3-031-16947-2_4
36. Tabansky, L., Lichterman, E.: PROGRESS: the sectoral approach to cyber resilience. Int. J. Inf. Secur. **24** (2025). https://doi.org/10.1007/s10207-024-00910-3
37. Tsai, C.H., Zdravkovic, J.: A survey of roles and responsibilities in digital business ecosystems. In: IFIP WG 8.1 Working Conference on the Practice of Enterprise Modelling (2020)

AI Auditing: Towards a Practicable Model

A. B. van Wingerden(✉) and H. Weigand

Tilburg University, P.O. Box 90153, 5000 LE Tilburg, The Netherlands
`bran.van.wingerden@gmail.com, h.weigand@uvt.nl`

Abstract. With the advancement of Artificial Intelligence (AI) technologies in the recent years, the business application of AI models has expanded significantly. Hence, auditors are progressively encountering AI systems, models and algorithms during audit and assurance projects. The growing scientific domains of eXplainable AI (XAI) and Responsible AI raise concerns around the transparency, explainability, and other ethicalities. These concerns, in combination with upcoming legislation, demand audit statements on reliability, integrity, and other aspects of AI models. Where auditing is well-established, AI auditing remains a novel practice. This research includes literature research, exploration of AI audit cases, and interviews with AI experts to discover relevant methods and specificalities of AI audits. Through the methodology of design science, a first structured AI Audit Process is developed and proposed to provide AI auditors with a flexible reference frame to conduct customised AI audits. This research is a step towards the advancement of an AI auditing method and offers valuable insights for science and practice.

Keywords: Artificial Intelligence · Auditing · Explainability · AI Auditing · Algorithm assurance

1 Introduction

In the last decade, the field of auditing changed significantly [1]. The increasing demand for real-time information led to auditing 3.0, characterized by the integration of (Big) Data Analytics [2]. Today, this profession stands at the brink of auditing 4.0, where Artificial Intelligence (AI) technologies are both the object of focus during audits and instruments for automating auditing activities. This shift in the domain of auditing seems imminent now, especially after OpenAI's release of ChatGPT in November 2022, which made impact on general awareness and jobs. Leading firms, like the Big 4 Accounting & Consulting firms, are now actively innovating, researching opportunities to maintain a competitive edge, and were forced to update their policies [3]. Despite the growing implementation of AI in audits, there is a notable lack of empirical research and standardized frameworks for auditing AI systems, particularly around issues of transparency and explainability. Our research addresses the question: '*How can AI auditors assure transparency and explainability in the practice of auditing AI systems?*'. The structure of this paper is as follows. In Sect. 2 we briefly describe eXplainable AI and AI auditing.

Section 3 outlines our research methodology, used to develop an AI auditing artefact. Section 4 presents our findings and the resulting artefact. Section 5 discusses the findings and limitations. Lastly, Sect. 6 draws our conclusions.

2 Background

Responsible AI (RAI) is an approach to developing and deploying artificial intelligence (AI) from an ethical standpoint. Arrieta [4] defines six foundational ethical principles, such as accountability, and fairness. Notably, there is no uniformity on definitions and concepts. Nor is there a universal accepted framework on RAI and the AI aspects [4] that could or should be audited [5]. The EU commission emphasizes trustworthy AI principles; fairness, human autonomy, prevention of harm, and explicability (explainability), as fundamental human rights [6]. Some of the RAI aspects can be interpreted through the ART framework [7]. **A**ccountability involves verifiability, traceability, and ethical trade-offs [4]. **R**esponsibility considers the interaction between human and AI and the role of all stakeholders [6]. **T**ransparency is about communication and will be elaborated below.

2.1 Explainable AI

RAI requires further research, but organisations can start by adopting its principles. Critically, RAI addresses AI that *"...does not yet exist"*, yet it could be considered as AI safety research for the eventuality of Artificial General Intelligence (AGI) [8]. eXplainable AI (XAI) is the object of a growing body of regulations and aims to provide meaningful, understandable and trustworthy outputs of AI systems [4]. AI models are categorised as '*black-box*' (opaque), '*white-box*' (interpretable but less accurate), or '*grey-box*' (balancing accuracy and transparency) [9, 10]. Complex AI systems (black-boxes) pose challenges in transparency and accountability. Robustness, adversarial vulnerability, and data quality further complicate XAI implementation [11]. The goal of XAI is to make AI systems and their outputs more understandable to humans [12]. The field of XAI distinguishes two types of models: Ante Hoc (interpretable) and Post-Hoc (explained through external XAI techniques) [4]. We adopt the following definitions: *Interpretability* is about understanding the model's internal workings so that provided insights can make sense for the targeted audience [9]. *Explainability* provides insights in the model's decisions in a human-readable format for end-users [9]. Both explainability and interpretability are pivotal in verifying ethical aspects of an algorithm [4].

Evaluation of Explainability
The literature offers a few frameworks to assess explainability. Nauta [12] proposes a framework, based on approximately 300 studies, identifying 12 key properties like correctness, completeness, and consistency e.g., each assessed via specific evaluation methods. Another study warns that an explanation might be as complex as the models themselves [9]. New quantitative metrics such as DoX [13] are under development. More established techniques are LIME and SHAP. Though useful, these methods require deep model understanding.

Evaluation of Transparency

Transparency is openness about the design of the AI artefact. This includes the model structure, the development process, the training data, the testing, among others [14, 15]. Transparency is directed to certain stakeholders, so basically it is a communication quality [16]. Transparency can be provided by making the system explainable, in one way or another, but this is not the only way and not always the best way. The design can also be communicated clearly by means of AI governance, such as mandated by regulation. This includes for instance openness about the datasets that have been used for the training and how biases were avoided or removed [6]. In conclusion, transparency is two-fold. Transparency can be provided through explanations, or it can be established and controlled through AI governance. AI transparency can be categorised in 4 areas to assess [7]. A checklist is not sufficient, more research is needed to define and measure AI transparency.

2.2 AI Auditing

Auditing is a professional and well-established practice [18]. IT audits assess confidentiality, integrity and availability and alignment with standards and organisational goals. Control testing evaluates whether internal controls cover risks. Through the lens of an auditor, an AI or algorithm introduces considerable digital or algorithmic risks that could result into real-world consequences [19]. Compliance, particularly under the 2024 EU AI Act, mandates explainability and transparency. The AI audit can be about several aspects, such as fairness or the question of whether the system is transparent enough to its customers, but doing an AI audit requires a minimum level of transparency towards auditing (auditability) in the first place [4]. In addition, data quality is crucial for AI audits. Data explainability enhances transparency by clarifying data origins, datasets of AI models' design, its development, the training, preprocessing procedures, biases, and model features. [9]. While often implicitly reported, data explainability remains to become an established concept. A concept in the field of IT auditing that is becoming more established is algorithm assurance, which is about risk management and control of 'risky algorithms' applied in AI products [19]. AI audits could follow traditional IT auditing steps [18]. Key auditable parts of algorithms are the data, the model, and the development process. The model may be considered a black-box. There are 3 algorithm assurance risk areas: Autonomy (human oversight), Complexity (technological and operational intricacies), and Influence (the scope and impact of an AI's decisions) [19]. Algorithm assurance – or AI auditing – has significant overlap with the principles from XAI and RAI. However, it is much wider, solely auditing the algorithm (or model) is markedly insufficient [19]. Legislation like the Digital Services Act (DSA), the Digital Market Act (DMA), and the EU AI Act (EUAIA) provide audit guidelines. DSA and DMA mandate transparency, risk assessments, and fairness in AI, especially for recommender systems. As of June 2024, the EUAIA serves as a reference frame, as its impact is still unfolding. Being compliant is part of an organisation's governance. Ethical principles must translate into practical governance. IT-, Data-, and AI-Governance frameworks support audits and offer reference for assessing alignment with ethical and regulatory standards [20].

3 Methodology

This study employed an exploratory design due to the nascent nature of the research areas XAI, RAI, and AI auditing. Design Science Research (DSR) is used as the theoretical framework, enabling the development of an artefact to address practical problems within the AI auditing domain [21]. The created artefact aims to provide guidelines to approach auditing AI-systems based on theoretical insights and practical applicability. While DSR involves multiple iterations, this research paper conducts a single iteration due to time constraints and resources at the research environment. Instead, an 'iteration' from preliminary model to actual artefact has been made. Our research combines literature studies, information from client cases provided by KPMG Netherlands, and expert interviews in order to develop and refine the artefact. Figure 1 showcases the exact steps undertaken during the research time, which is conducted between February and June 2024. The process, schematically presented, is based on the DSR theory [22] and its established process [23].

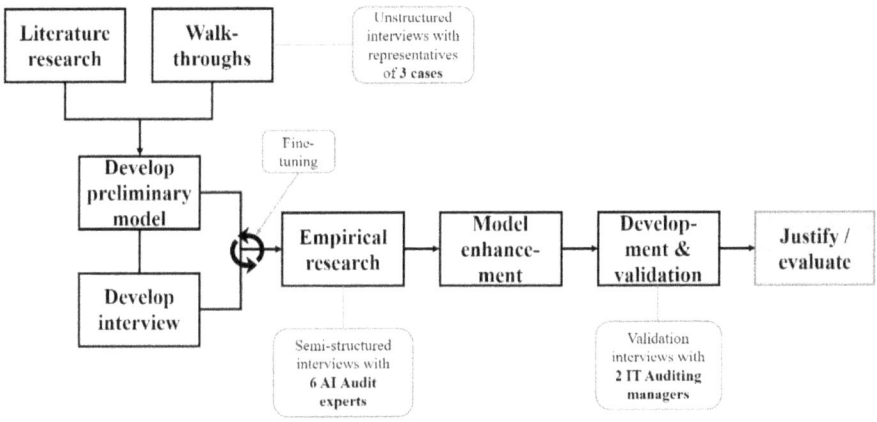

Fig. 1. Research design.

The literature study established a foundation and baseline of reasoning. The keywords for the literature research were *transparency, explainability, AI ethics, Responsible AI, AI systems, AI auditing, auditing, AI assurance,* and *explainable AI*. For the Case Walkthroughs, unstructured interviews with representatives from three anonymised KPMG clients (Case Organisation (CO) CO1, CO2, CO3) were conducted to explore AI audit challenges and gather practical insights (Table 1).

Note that using case information, such as documentation and unstructured interviews with people involved, are dissimilar to conducting a case study. A case study requires an in-depth and highly detailed examination of a case with narrow focus [24] and we did not aim for that. In DSR, the cases should be seen as possible **use cases** of the artefact. The research made use of expert interviews, of a semi-structured nature, with AI audit experts in consultant, audit, manager or director roles. The interviews provided in-depth perspectives. A total of six interviews were conducted (Table 2.). Additionally,

Table 1. Case organisations.

Organisation reference	Type of organisation
CO1	Major Dutch insurance company
CO2	Multinational media Platform (VLOP)
CO3	Multinational e-commerce company (VLOP)

2 validation interviews were conducted to gather feedback from two IT audit managers in order to refine the artefact with a general audit perspective. The interviews were analysed preceding the next interview in order to fine-tune the data collection method. The nature of DSR allows for adjustments during the data processing and collection phase [21]. Due to confidentiality and limited access to client information, this study relied on high-level case insights rather than detailed case studies

Table 2. Interviews & validation interviews.

Interviewee	Role	Interviewee	Role
Interviewee A	Snr. RAI Manager	**Interviewee E**	Snr. Consultant
Interviewee B	RAI consultant	**Interviewee F**	Director
Interviewee C	Snr. RAI consultant	(validating) **G**	Manager
Interviewee D	RAI consultant	(validating) **H**	Manager

4 Results

This chapter outlines the results and findings from the literature and interviews for the creation of an artefact for AI auditing. The analysis transforms insights from the literature, case walkthroughs, and interviews into requirements for the artefact. The case walkthroughs are used to formulate, together with the literature, a preliminary model.

4.1 Preliminary Model

The preliminary model draws its aspects from literature findings and the initial walkthroughs. From the literature, inter alia, two audit contexts can be distinguished [25]. In the **Make Context** the audit focuses on the AI model and its development environment, including design, testing, and controls. CO1 is an example of a detailed investigation of an AI model's reliability. The **Use Context** on the other hand focuses on the AI system's performance within a business context, with its procedures and processes. Audits at CO2 and CO3 put the focus on the AI's integration in complex processes. The different contexts in different projects suggest that there are 'types' of audits. Hence, a preliminary typology was created (Fig. 2), a first step in the DSR process.

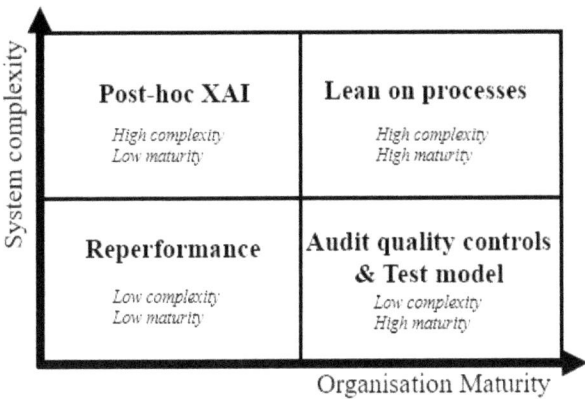

Fig. 2. AI audit typology

The model takes the form of a quadrant defined by two axes. The vertical axis represents 'System Complexity' which reflects the degree to which an AI system is a 'black-box' or hard to investigate. It is likely that a multi-layered ANN is harder to investigate than a ML model with a clear set of business rules. This was an important aspect derived from the literature. The horizontal axis represents 'Organisation Maturity' and represents the structure, processes, and controls surrounding AI systems. The walkthroughs made clear that there is a difference in how the Case Organisations (CO's) were organised around their models. Organisation maturity has two aspects: The *presence of controls* and *organisational predictability*. The more predictable a business process the more it could be considered mature.

4.2 Requirements from the Interviews

The findings from six interviews with AI (audit) experts returned relevant aspects and insights for an AI auditing artefact. Interviewees were all from KPMG's Responsible AI (RAI) department, experienced in (AI) model validations and AI risks (Table 3).

4.3 The AI Audit Process

From the requirements that are based on literature, case walkthroughs, and interviews, a stepwise approach, or process, to AI auditing was formulated (Fig. 3). The presented artefact is still far from comprehensive, but aims to reflect the current best practices in AI auditing in the form of a practicable process or tool.

The AI audit process in Fig. 3 is divided in overarching phases. The first phase involves iterative steps, based on CRISP-DM, to build an understanding of the AI audit context, including business processes, data, and models. Besides the understanding of the processes, alignment with organisational goals, regulations and the datasets used by the model, another step is added. Model understanding aims to comprehend the AI model's characteristics in the business setting, judged on the algorithm risks [19]. After the understanding an auditor identifies specific risks associated with the situation. The

Table 3. Requirement aspects from interviews. IntRef refers to the interviewee(s) addressing the aspect.

Empirical artefact aspects	Requirement	IntRef	Case reference
Intake	Every AI audit must start with an initial exploration to investigate the scope and objective	A, B, C, D, E	CO2
Audit objective	Like regular audits the objective and purpose need to be determined	A, C, E	CO1
Audit entity	An AI audit must determine and scope the audit entity; the AI model, the processes, the controls	A, E	CO1
Maturity	The maturity and presence of the processes and controls must be assessed	A, B, C, D	–
Complexity	AI audits must determine whether the model can be investigated or should be considered black-box	A, B, C, D, E	–
Continuous evaluation	AI auditors must always evaluate what worked and what did not work at the end of an AI audit project	A, D, E	–
Professional judgement	The model must allow for professionals to apply judgement from their own expertise	A, B D, F	CO1, CO3
Limited standardisation	All AI audits encounter different elements, requiring the AI audit to be flexible on the detailed levels	A, D, F	CO1
Ethical risk framework	Every AI audit has its own risks and ethical concerns and must make a specific selection of which are applicable for the case	A, F	CO1
Regulatory compliance	The AI audit must be aligned with compliance audits such as DSA / DMA	B, C, D	CO2, CO3

(*continued*)

Table 3. (*continued*)

Empirical artefact aspects	Requirement	IntRef	Case reference
Controls *(or pillars of control)*	5 pillars; *reliability, resilience, explainability, accountability,* and *fairness* must always be audited	A, C, F	CO1, CO2
Segregation of duties	Every AI audit should investigate administrative separation of roles and others involved in the context	A, F	CO1, CO2
CRISP-DM	The initial phases of CRISP-DM provide understanding, which is required for every individual AI audit case	C, *(A, B, D)*	CO2
AI-Lifecycle	Ethical considerations need to be incorporated throughout the lifecycle	A, B, C, D, E	-

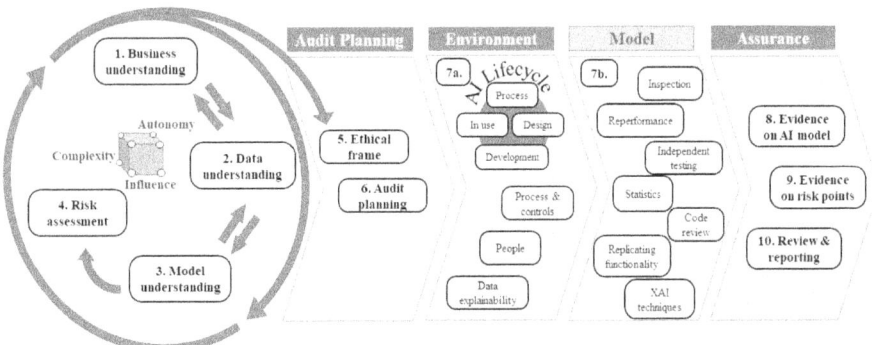

Fig. 3. The AI audit process (initial DSR artifact)

risk profiles vary depending on the nature of the AI audit, as seen in the cases. The audit planning phase determines the scope of the AI audit and determines whether the audit is focused on the Make or Use context, or both. Preceding to the planning, an ethical scope is defined through the selection of risks and ethicalities, specific to the AI audit. This step is not to set ethical standards, but rather to tailor the audit to specific concerns.

The seventh AI audit process step is divided by the Make context and Use context, or in simplified terms; the Environment and the Model itself. The *Environment* phase evaluates controls and monitoring mechanisms around the AI model, whereas the *Model* phase directly examines the model. An auditor could opt to either investigate one or both. Lastly, the assurance phase is concentrated on gathering evidence and presenting the audit findings. When the evidence gathered from the AI model audit (step 7b), the auditor may revisit environmental control for additional coverage of risk points. For that reason, step 8 precedes step 9. When all controls, from the *Environment* or *Model*, cover the risk point sufficiently, the evidence can be considered adequate and can be compiled into an assurance report, typical to IT audits.

5 Discussion

Our research explored the question how AI auditors can assure transparency and explainability in the practice of auditing AI-systems by combining insights from XAI theory and empirical findings. The research highlights a nuanced difference between Transparency and explainability. Transparency involves disclosing the design, while explainability ensures model outputs are interpretable. Addressing the complexity of AI systems without standardised methods is a big challenge for AI auditors. The developed artefact offers an initial framework for AI audits, bridging theory and practice. Empirical findings indicate that the mandated transparency is a prerequisite for audit investigation, while explainability is not specifically audited or assured and is mainly used as a 'tool' to gain clarity on model outputs. Interviews confirm current audits rely on organisational statements rather than external (legal) standards or frameworks. At times, internal claims were validated. However, in other instances audits revealed regulatory ambiguities. Concluding, transparency varies across contexts. As a result, AI audits assess organisational statements against documentation, internal controls, processes, the development cycle, or AI entities in a more generalised way, following the known general IT audit structure. The XAI domain is still evolving, quantitative metrics remain unsuitable since they lack a standardised benchmark, if that would make sense at all. Logically, the process is audited more than the technology. The AI Auditing Process artefact that we have developed still lacks DSR justification, as practical case studies were not feasible for refinement through real-world application. Instead, validation interviews were held. Through these interviews the artefact was theoretically validated and recognised as an (AI) audit process. Furthermore, the findings reflect some company bias. Although the (use) case organisations are either multinational or operating on the international market, the study was conducted solely within KPMG Netherlands.

6 Conclusion

The rapid evolvement of AI fundamentally reshaped auditing, necessitating new approaches to transparency and explainability, as defined in XAI. Despite AI's advancements the auditing principles – ensuring accuracy, reliability, completeness, integrity, accountability, correctness, and ethicalities – remain unchanged. The AI Auditing Process artefact extends traditional methods, providing auditors with a structured approach to potentially enhance assurance and trust. A result of a combination of literature research, use case explorations, and interviews. XAI views transparency as an enabler for AI audits. Explainability encompasses the broader context and serves more as a tool. Future research should test and refine this artefact through practical (use) case studies and longitudinal studies to assess whether structured AI audits foster trust. Further investigation to differentiate transparency, explainability, and interpretability in practice is also needed. This research contributes to the growing body of knowledge on AI auditing, algorithm assurance and XAI. Nonetheless, sustained collaboration between researchers, AI auditors, and public administration is imperative to effectively shape the future of AI auditing.

References

1. Lombardi, D.R., Bloch, R., Vasarhelyi, M.A.: The future of audit. J. Inf. Syst. Technol. Manag. **11**(1) (2014). https://doi.org/10.4301/s1807-17752014000100002
2. Alles, M.G., Dai, J., Vasarhelyi, M.A.: Reporting 4.0: business reporting for the age of mass customization. J. Emerg. Technol. Account. **18**(1), 1–15 (2021). https://doi.org/10.2308/jeta-10764
3. Almufadda, G., Almezeini, N.A.: Artificial intelligence applications in the auditing profession: a literature review. J. Emerg. Technol. Account. **19**(2), 29–42 (2021). https://doi.org/10.2308/jeta-2020-083
4. Arrieta, A.B., et al.: Explainable Artificial Intelligence (XAI): concepts, taxonomies, opportunities and challenges toward responsible AI. Inf. Fusion **58**, 82–115 (2020). https://doi.org/10.1016/j.inus.2019.12.012
5. Laine, J., Minkkinen, M., Mäntymäki, M.: Ethics-based AI auditing: a systematic literature review on conceptualizations of ethical principles and knowledge contributions to stakeholders. Inf. Manag. **61**(5) (2024). https://doi.org/10.1016/j.im.2024.103969
6. HLEG AI.: Ethics Guidelines for Trustworthy AI. European Commission (2019). https://ec.europa.eu/futurium/en/ai-alliance-consultation.1.html
7. Dignum, V.: Artificial Intelligence: Foundations, Theory, and Algorithms. Springer, Heidelberg (2019). https://doi.org/10.1007/978-3-030-30371-6
8. Tegmark, M.: Life 3.0: Being Human in the Age of Artificial Intelligence, 2nd edn. Penguin Books, London (2018)
9. Ali, S., et al.: Explainable artificial intelligence (XAI): what we know and what is left to attain trustworthy artificial intelligence. Inf. Fusion **99**, 101805 (2023). https://doi.org/10.1016/j.inffus.2023.101805
10. Casper, S., et al.: Black-box access is insufficient for rigorous AI audits (2024). https://doi.org/10.1145/3630106.3659037
11. Weber, L., Lapuschkin, S., Binder, A., Samek, W.: Beyond explaining: opportunities and challenges of XAI-based model improvement. Inf. Fusion **92**, 154–176 (2023). https://doi.org/10.1016/j.inffus.2022.11.013
12. Nauta, M., et al.: From anecdotal evidence to quantitative evaluation methods: a systematic review on evaluating explainable AI. ACM Comput. Surv. **55**(13s), 1–42 (2023). https://doi.org/10.1145/3583558
13. Sovrano, F., Vitali, F.: An objective metric for explainable AI. Knowl.-Based Syst. **278**, 110866 (2023). https://doi.org/10.1016/j.knosys.2023.110866
14. Laato, S., Tiainen, M., Najmul Islam, A.K.M., Mäntymäki, M.: How to explain AI systems to end users: a systematic literature review and research agenda. Internet Res. **32**(7), 1–31 (2022). https://doi.org/10.1108/intr-08-2021-0600
15. Schmidt, P., Biessmann, F., Teubner, T.: Transparency and trust in artificial intelligence systems. J. Decis. Syst. **29**(4), 260–278 (2020). https://doi.org/10.1080/12460125.2020.1819094

16. Rendon, L.G.: An introduction to the principle of transparency in automated decision-making systems. In: 2022 45th Jubilee International Convention on Information, Communication and Electronic Technology (MIPRO), pp. 1245–1252 (2022). https://doi.org/10.23919/MIPRO55190.2022.9803417
17. Balasubramaniam, N., Kauppinen, M., Rannisto, A., Hiekkanen, K., Kujala, S.: Transparency and explainability of AI systems: from ethical guidelines to requirements. Inf. Softw. Technol. **159**, 107197 (2023). https://doi.org/10.1016/j.infsof.2023.107197
18. Rittenberg, L.E., Johnstone, K.M., Gramling, A.A.: Auditing: A Business Risk Approach, 7th edn. South-Western College Pub. (2010)
19. Boer, A., de Beer, L., van Praat, F.: Algorithm assurance: auditing applications of artificial intelligence. In: Progress in IS, Part F2545, pp. 149–183. Springer Medizin (2023). https://doi.org/10.1007/978-3-031-11089-4_7
20. Mäntymäki, M., Minkkinen, M., Birkstedt, T., Viljanen, M.: Defining organizational AI governance. AI Ethics **2**(4), 603–609 (2022). https://doi.org/10.1007/s43681-022-00143-x
21. Wieringa, R.J.: Design Science Methodology for Information Systems and Software Engineering. Springer, Heidelberg (2014). https://doi.org/10.1007/978-3-662-43839-8
22. Hevner, A., March, S., Park, J., Ram, S.: Design science in information systems research. MIS Q. **28**(1), 75 (2004). https://doi.org/10.2307/25148625
23. Johannesson, P., Perjons, E.: An Introduction to Design Science. Springer, Heidelberg (2014). https://doi.org/10.1007/978-3-319-10632-8
24. Yin, R.K.: The case study crisis: some answers. Adm. Sci. Q. **26**(1), 58 (1981). https://doi.org/10.2307/2392599
25. Weigand, H., Johannesson, P., Andersson, B.: An artifact ontology for design science research. Data Knowl. Eng. **133**, 19 (2021). https://doi.org/10.1016/j.datak.2021.101878

From Acquiring to Suggesting DL Design Choices with Agility: A System Design

Gustavo Rodrigues dos Reis[1,2(✉)], Mario Cortes Cornax[1], Adrian Mos[2], and Cyril Labbé[1]

[1] Univ. Grenoble Alpes, CNRS, Inria, Grenoble INP, LIG, 700 Av. Centrale, 38401 Saint-Martin-d'Hères, France
{mario.cortes-cornax,cyril.labbe}@univ-grenoble-alpes.fr
[2] NAVER LABS Europe, 6 Chemin de Maupertuis, 38240 Meylan, France
{gustavo.rodrigues-dosreis,adrian.mos}@naverlabs.com

Abstract. Modeling new applications with deep learning (DL) algorithms requires substantial knowledge. Some systems aim to simplify design choices by providing support for specific pre-defined use cases, like blurred image backgrounds or text summaries, making it easier by limiting certain options. There is a gap in addressing diverse use cases and efficiently gathering knowledge output from the deep learning community to find and reuse models and datasets from various sources if they help solve a use case. In this experience study, we are interested in how to suggest and manage DL design choices stemming from artifacts published by the DL community to help non-expert users. We detail a system for this end using a business process (BP) model, discussing the requirements for software components implementing each BP model task. We also analyzed agility in recomposing pipelines using an in-house tool against open-sourced orchestration tools, implementing deep learning model adaptation components in one highly modular BP model task.

Keywords: Process modeling · User Guidance · AI · Deep Learning

1 Introduction

Deep learning models have gained popularity since they demonstrated their success in several applications in this past decade of development. Two main factors were responsible for this growth: data availability, as many more sensors and media data are uploaded daily to the web [7], and scalability since popular neural network architectures are highly parallelizable using massively parallel processor (MPP) hardware [12] (GPU, TPU, FPGA, etc.) available today. After the challenge of cleaning raw data and filtering out noise [10], the next one is making effective design choices for the deep learning model to extract useful patterns.

This challenge is partially due to the data-dependent nature of those algorithmic design choices, which are commonly approached *ad hoc*. The preference is given to the most popular DL networks, that score higher on benchmarks and have frameworks built around them as transformers and torchvision. Basing deep learning design choices solely on those factors can hinder discovering more effective or efficient existing choices for specific needs [3].

The path of trial and error in data-driven AI methods could benefit from the knowledge that stemmed from successfully implemented design choices of deep learning methods in the past. The context descriptions in which those applications were designed might allow a system to look for past experimentation artifacts and find candidates of suitable choices for new applications.

In this work, we define an approach to a Business Process Model (now called the BP model) to guide novices through different design choices for DL models (e.g., model architecture, hyperparameter choice) in an agile way. This approach is a system architecture to help tailor deep learning choices to a user. Its core principle is to reuse knowledge from training artifacts released by the deep learning community.

Throughout the text, we exemplify concepts and processes for the specific use case of tagging photos to the locations where they were taken. In the deep learning community, this intent is named Visual Place Recognition (VPR) [13].

This work is structured as follows: Sect. 2 outlines characteristics of existing systems that assist non-experts in adapting deep learning models, highlighting the differences from our approach. Section 3 provides definitions and notations used throughout the paper. Section 4 details the proposed method's architecture via a BPMN process model, focusing on two main parts: *Knowledge Registration* and *Recommend Parametrization of DL Design*, including tasks and their dependencies. Section 5 discuss latency requirements related to DL pipeline's volatility, accompanied by an analysis of orchestration tools for Machine Learning. Finally, Sect. 6 offers concluding remarks.

2 Related Works

Helping less tech-savvy users operate a new technology happens during the later stages of *innovation diffusion theory* [15], when technology gets simplified. The current mechanisms that address this simplification need for *deep learning modeling* intervene only on *final use cases*, packaging DL models in pre-defined services, restraining or removing all choices in the network design from the user and letting them control mostly the input-output schemes definition.

Examples of such systems can be found in Teachable Machines and Tensorflow Playground [5,11]. Other systems, such as Azure ML, Sagemaker AI and Vertex AI, operate mainly with Deep Learning as a Service (DLaaS), where vendors add more controllability to certain aspects of the neural network design but hide the models and most parameters behind them to focus on the service provided.

There is value in documenting experimentation when a technological practice operates on knowledge that depends highly on *ad hoc* experimentation. The availability of this information can be used to guide the next practices. Standards for documentation and logging experiments such those in Health Level

Seven Standards(HL7) in healthcare data treatment and NIST Cybersecurity Framework (CSF) help knowledge reuse and guide practice in their respective fields.

In deep learning, several ways exist to share a developed model's training artifacts publicly, but no widely followed standard exists. Typically, researchers share novel models or techniques through anonymized software repositories that include relevant information and files. Also, models can be distributed in software libraries aiming for a common end goal or data input modality. Another common method of sharing models and corresponding artifacts is through model hubs, such as those found in Hugging Face (HF) and Kaggle. HF, in particular, has gained traction due to its integration with the *transformers* python library.

It is also worth mentioning that the transition to using hubs pushed the development of standards for dataset metadata [1]. Their adoption has gained traction, but the datasets that implement them are still a small fraction of the overall published in those hubs.

The diversity of ways datasets and models are distributed mandates to be adaptable when using them for discovering best practices for new cases. The modular and robust approaches for building software pipelines with DL models create an abstraction layer for training artifacts while orchestrating containers [6, 8]. However, they are not agile enough to switch to different datasets and models in an iterative exploratory setting that we need for reusing DL artifacts available. This drawback produces the need for a methodology that can be modeled with less rigid tooling, which we detail a proposition later in this work, with the next section grounding important concepts.

3 Used Notation and Concept Definitions

This section outlines the notations and definitions for tasks related to managing and searching deep learning (DL) model design choices based on user requirements, along with relevant concepts in deep learning practice. The following list compiles the notation choices used throughout the text:

- **Training Artifact**: Refers to any byproduct produced during the training phase of a deep learning model, as noted in [2]. This includes code for the model architecture, data processing for the model format, training iterations, and files with trained model parameters or hyperparameters. In the VPR example, it might include a model architecture script and its documentation.
- **Dataset Store**: A system for efficiently managing the storage and retrieval of datasets. It uses a data store repository model for dataset objects, denoted as $\mathcal{D}_{data}$. For image classification tasks like VPR, the dataset store must include datasets with photos of places taken from various angles and under different lighting conditions to support place identification.
- **Model Store**: A system for managing the storage and retrieval of deep learning models and their training artifacts, denoted as $\mathcal{D}_{models}$. For the VPR example, it must include DL model artifacts generated from datasets used to identify place tags in images.

- **Task** (business process model context): an atomic activity within a process flow [16]. It is referred to throughout the text by the notation $TaskI$(inputs), in which $TaskI$ is an initialism of the task's name.
- **DL Task**: A defined input-output scheme, including their semantics and the machine-learning paradigm to derive outputs. For instance, in place identification of images: (*Input: Image, Output: Tag, MLProblem: Classification*), where Tag can represent geographical coordinates or a name.
- **Stored Item**: any record available in the previously defined stores. We reference it in the text as id while the contents of the item pointed through this reference are referred to as d_{id}.

The definitions of tasks in the business process model described later on also make use of the following concepts usually present in machine learning contexts [9]:

- **Sample**: a single piece of data in a dataset from which a deep learning model generates an output. Label is the metadata associated with a sample, indicating what the model should output. For instance, an image of a place with a label serves as one sample in the VPR example.
- **Machine Learning (ML) problem**: a particular machine learning setting with a certain ML model's output and input conditions. A non-exhaustive list of examples of the issues solved in ML theory, referred to as an ML problem, are classification, regression, and clustering, [4].
- **Curriculum Learning**: subfield of machine learning focused on techniques that influence the order of training samples based on difficulty. This approach is analogous to educational curricula. For instance, in our VPR example, varying light conditions can define a curriculum that progresses from easier to more challenging samples for correct classification.
- **Learnable/Model Parameters**: the set of optimized numeric parameters that are subject to be updated during a training run of a DL model.
- **Hyperparameters**: Configuration parameters set before training a deep learning model that remain constant during optimization. They can be adjusted during hyperparameter optimization (HPO). Examples include the number of layers in a neural network, input and output dimensions, and optimizer parameters.

4 System for Suggesting DL Design Choices from Existing Artifacts

The method defined here for suggesting DL design choices from existing artifacts is illustrated here with a business process model. Guiding users through deep learning algorithm with this design is based on two distinctive parts. The first one concerns leveraging latent knowledge in the DL training artifacts to help users, namely *"Knowledge Registration."* The second is consulting this knowledge to attend to the user's use case, *"Recommend Parametrization of DL Design."* We introduce each and the respective details for the constituent tasks below.

4.1 Knowledge Registration

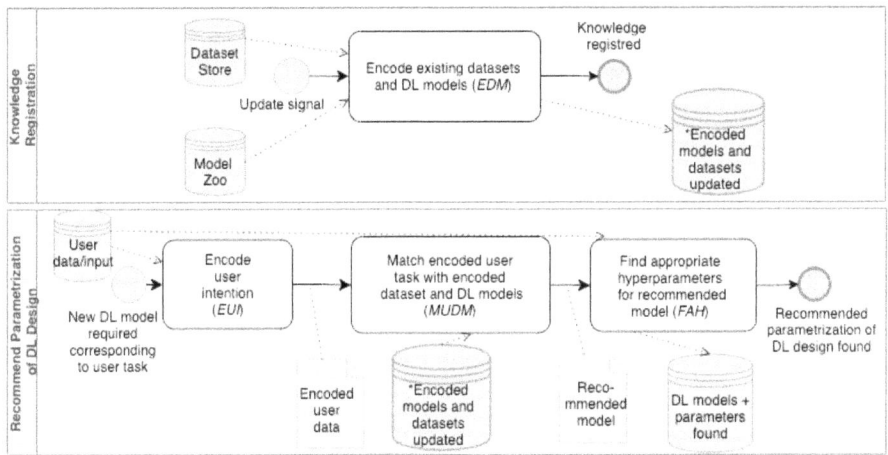

Fig. 1. BPMN model representing the two main parts of our method, namely *Knowledge Registration* that encodes artifacts in the format our system can consult, and *Recommend Parametrization of DL Design*, that help to find good DL choices for a given use case

Our proposition is based on routing users' use case DL requirements to related past DL designs published in repository sources. Therefore, we prioritize establishing a process in which important decisions on each DL application on those sources are identified and registered appropriately.

The primary task in this aspect within the illustration is incorporating the artifacts from the target repositories into the system knowledge base. This knowledge base comprises two stores: the dataset store and the model store (also called model zoo).

This task in the process model is named *"encode existing datasets and DL models,"* as indicated in Fig. 1

Encode existing *datasets* and DL *models* (EDM) Encoding in this context refers to creating data representations of pertinent parts of DL training artifacts, allowing their reuse. The encoding procedure for datasets and the one for models focus on the following aspects of each:

Datasets: in our approach, the relevant dataset elements are a sample's content and the label(s) associated with it, if any.

Models: the relevant model elements to be identified allow the instantiation of a model. This implies any code or API reference from a library containing the different parts of the model. A discussion on all EDM task underpinnings, but also for the other tasks in our system, is beyond the scope of the paper. Briefly, it receives datasets and model artifacts in the kind of stores described in the previous section as inputs and output entries for the system knowledge store encoded accordingly to the respective artifact type, $\mathcal{D}_{knowledge} = EDM(\mathcal{D}_{models}, \mathcal{D}_{data})$. For the VPR example, the system EDM task previously organized the information of datasets and models of various use cases so the system's following tasks could search through them, identify and prepare only the ones most relevant to place identification in images.

4.2 Recommend Parametrization of DL Design

In this subsection, we present the tasks involving the path through understanding what the user consulting the system expects to do and searching for compatible design choices for a solution. This search is possible because the knowledge store previously constructed using artifacts collected from the community, details of which were discussed in the previous subsection.

There are three tasks, each dependent on its predecessors' results, meaning that the triggering signals are the output information from a previous activity.

Encode User Interaction (EUI). The first task in Fig. 1 serves as the entry point after users make a request. It takes user data, or D_{user}, as input and prompts for additional information to clarify the goal of the deep learning design. The final output describes this intention as $(\mathcal{I}, \mathcal{O}, MLProblem) = EUI(D_{user})$, where $\mathcal{I}$ refers to components of D_{user} input to the neural network and $\mathcal{O}$ denotes the desired output types.

User interaction may be needed when the input elements for the DL model can't be determined solely from the data file structure. Similarly, if there's uncertainty about the ML problem based on data annotations, the user will be prompted for additional information. After this, a summary of the DL design requirements is produced. For our example, this means that we can describe the problem the user is trying to solve as the specification present in Sect. 3: (*Input: Image, Output: Tag, MLProblem: Classification*).

Match Encoded User Specification with Encoded Dataset and DL Model (MUDM). The second task in Fig. 1 is to "*match encoded user specification with encoded dataset and DL model*" (MUDM). It uses the summary from $EUI(.)$ and user data to provide guidance for DL design. This task analyzes the structure and content of user data to find similar datasets previously used to train DL models. Identifying these similarities connects user needs with DL knowledge, helping to inform model design suggestions.

The output of this task is a model reference in the model store or a composition of models, including the code and parameters to instantiate a known deep learning approach. In the VPR example, this involves a model for image feature extraction and a classifier that processes the output.

Find Appropriate Hyperparameters for the Recommended Model (FAH). The third task, after MUDM, is to "find appropriate hyperparameters of the recommended model" (FAH). Taking the model reference as input, this task searches for good hyperparameters to give the user a more personalized suggestion of DL model design choices. It aims for efficiency in the hyperparameter optimization (HPO) process by characterizing the difficulty of individual user data samples, either automatically or with user input [14]. This approach allows HPO to proceed while applying curriculum learning principles. The output is a model reference with a new, personalized parameterization.

5 Discussion on Agility

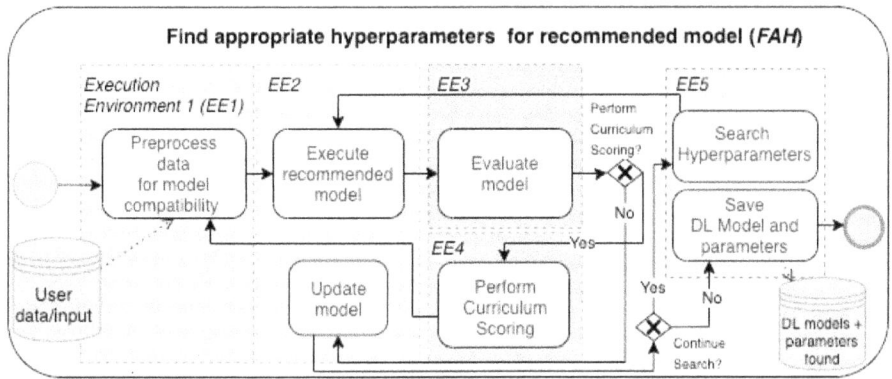

Fig. 2. A BPMN model representing tasks in a possible realization of pipeline within the FAH task. A rectangular dashed rectangle represents a software execution environment.

An important requirement for the proposed solution is that any instantiation of the business process model must offer agility in swapping the different models and options in a pipeline. This requirement arises from the system's iterative nature: several possible models and configurations must be run, which implies changing execution environments, to find a personalized solution for the user.

We investigated how agile can be the exchange of different components of a deep-learning pipeline appearing in a use case of our solution. This kind of pipeline is typically present in the execution logic of the FAH BPMN model task.

To select possible alternatives to implement the deep learning pipelines that appear in FAH and evaluate agility, we sought to understand the differences between open-sourced pipeline management systems by comparing those who have built large communities around them. We only keep those which repositories have at least 1,000 stars.

We are most interested in tools that *focus on machine learning pipelines, offer isolation for different components in this pipeline,* and *have confirmed*

Table 1. A comparative table of different open-source projects which pipeline orchestration is the goal or a feature.

	Kubeflow	Apache Airflow	Metaflow	MLRun	Polyaxon	Argo	Flyte	TFX	MLflow
ML Focused	✓	✗	✓	✓	✓	✗	✓	✓	✓
Container Oriented	✓	✗	✗	✓	✓	✓	✓	✗	✗
Popularity (in 10^3★)	14.6	38.4	8.5	1.5	3.6	15.3	5.9	2.1	19.3

support from an established user community. Table 1 summarizes characteristics that auxiliated our tool selection.

The comparison we establish in Table 1 shows for each tool the following characteristics: if they are planned for machine learning pipeline abstractions, and if containerization of software is the only form of execution isolation available. The popularity column shows the number of stars each tool received in its respective repositories.

Kubeflow and MLflow are the two main open-source alternatives used in industry to *design deep learning pipelines* that support separate execution environments regarding the number of stars in their respective repositories. We chose those two for further analysis as they are the *most popular ML-focused* options. Kubeflow is explicitly container-oriented for isolating execution environments, while MLflow, via its pipeline orchestration feature, makes it optional to use isolated environments to execute code components. This offers the possibility to choose between virtual environments and containers, and we use the former to diversify approaches.

The comparison is rooted in the agility for component change given a deep learning pipeline, translating to how long it takes to be ready to be executed again once one of its components is replaced. This change is expected within deep learning pipelines, as models and other data processing components can depend on different dependencies that demand different execution environments. The agility factor, in our case, is defined as the time elapsed between the command of deactivating a pipeline component and instantiating its replacement coupled to a different execution environment. This factor will be referred to as t_a.

We evaluate t_a of such tools compared to a custom one. Such a custom tool is a manager of isolated execution environments that can be reused if needed. We refer to it as an *"in-house tool."*

The deep learning pipeline used in the test presented and exemplified below in Fig. 2 is for a classification task, as the VPR example, since it is one of the simplest arrangements of DL pipeline possible. However, the task realized by the pipeline does not impact in t_a, and it is mentioned for contextualization. The t_a reported is only due to the tool's overhead in replacing any pipeline component with different execution environments.

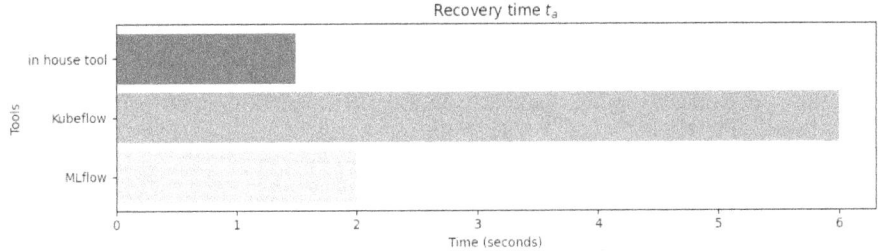

Fig. 3. The time elapsed between the replacement signal of a component in a pipeline that requires using a different environment until the replacement is ready to execute.

This pipeline shown in Fig. 2 has code components that execute the following tasks: "*Preprocess Data for Model Compatibility*," which transforms the data as the model expects to receive it; "*Model Execution*," in which the model is used to process the data, "*Evaluation*," in which some metric is computed to assess learning and performance between iterations, "*Model Update*," which compute new values for parameters of the model based on the previous results, "*Hyperparameter Search*," that applies the search for new configuration parameters for the model, and "*Curriculum Scoring*" in which the sequence of which sample in data comes can be decided, the default being random.

Among all code components in this pipeline, only "*Model Execution*" and "*Model Evaluation*" necessarily share the same execution environment since they host the execution of a given model. The latter is also more prone to change if we want to test different models that might depend on different frameworks or versions of a certain framework.

We compared t_a values for replacing the model component that requires a change of running environment in each of the three tools mentioned, Kubeflow, MLflow, and our *in house* environment manager. The results presented in Fig. 3 are statistically significant with an error margin of 5%, only considering the tool overhead to the processing of a replacement.

We can observe a greater overhead with Kubeflow, partially explained by its type-checking mechanism within a pipeline. This mechanism is important for the deployment of models but detrimental for processes expecting several changes in the pipeline throughout execution. The other open-source tool tested, MLflow, presents a better t_a for systems that would require a constant change in the pipeline, and it is expected as the pipeline orchestrating feature is one component of DL lifecycle management within MLflow, which is not as thorough as a standalone tool for DL pipelines.

The *in house tool* mentioned has a slightly better t_a than MLflow shows. It is not, however, part of a system specifically designed for deep learning but to manage execution environments of generic microservices. In all three scenarios, a reduction of order of magnitude is expected if we want to make the system described in Sect. 4 a responsive experience for a user, as we expect many components running on conflicting environments to be used and replaced.

6 Conclusion

In conclusion, this work addresses the challenge of knowledge accessibility in deep learning (DL) model design for non-expert users. Existing systems tend to focus on specific use cases, missing the broader spectrum of DL applications and community knowledge. We proposed a method for managing and suggesting DL design choices by leveraging knowledge from community-published training artifacts. Our approach is illustrated as a Business Process Model, detailing latency requirements for component switching. Our goal is to create a more accessible system for DL model development, empowering non-experts and facilitating the diffusion of this technology. Future plans include evaluating the method with complex user requirements and quantifying the quality of processed artifacts. Additionally, we aim to develop DL design suggestions for challenging edge use cases where model and data artifacts may be hard to find.

References

1. Akhtar, M., et al.: Croissant: a metadata format for ML-ready datasets. In: Proceedings of the Eighth Workshop on Data Management for End-to-End Machine Learning, DEEM 2024, pp. 1–6. ACM, New York (2024)
2. Aravantinos, V., Diehl, F.: Traceability of deep neural networks (2019). arXiv:1812.06744 [cs]
3. Bai, Y., Mei, J., Yuille, A.L., Xie, C.: Are transformers more robust than CNNs? In: Advances in Neural Information Processing Systems, vol. 34, pp. 26831–26843. Curran Associates, Inc. (2021)
4. Bishop, C.M.: Pattern Recognition and Machine Learning (Information Science and Statistics). Springer, 1 edn. (2007)
5. Carney, M., et al.: Teachable machine: approachable web-based tool for exploring machine learning classification. In: Extended Abstracts of the 2020 CHI Conference on Human Factors in Computing Systems, CHI EA 2020, pp. 1–8. ACM, New York (2020)
6. Chen, A., et al.: Developments in MLflow: a system to accelerate the machine learning lifecycle. In: Proceedings of the Fourth International Workshop on Data Management for End-to-End Machine Learning., DEEM 2020, pp. 1–4. Association for Computing Machinery, New York (2020)
7. Clissa, L., Lassnig, M., Rinaldi, L.: How big is Big Data? A comprehensive survey of data production, storage, and streaming in science and industry. Front. Big Data **6** (2023)
8. Gill, K.S., et al.: Utilization of Kubeflow for deploying machine learning models across several cloud providers. In: 2023 3rd International Conference on Smart Generation Computing, Communication and Networking (SMART GENCON), pp. 1–7 (Dec 2023)
9. Goodfellow, I., Bengio, Y., Courville, A.: Deep Learning. MIT Press (2016)

10. Goyle, K., Xie, Q., Goyle, V.: DataAssist: a machine learning approach to data cleaning and preparation. In: Arai, K. (ed.) Intelligent Systems and Applications, pp. 476–486. Springer Nature Switzerland, Cham (2024)
11. Hoeiness, H., Harstad, A., Friedland, G.: From tinkering to engineering: measurements in Tensorflow playground (2021). arXiv:2101.04141 [cs]
12. Li, S., et al.: Colossal-AI: a unified deep learning system for large-scale parallel training. In: Proceedings of the 52nd International Conference on Parallel Processing, ICPP 2023, pp. 766–775. ACM, New York (2023)
13. Lowry, S., et al.: Visual place recognition: a survey. IEEE Trans. Rob. **32**(1), 1–19 (2016)
14. dos Reis, G.R., Mos, A., Cornax, M.C., Labbé, C.: Data selection driven by item difficulty: on investigating data efficient practice for hyperparameter search. In: Proceedings of the IEEE/ACM 3rd International Conference on AI Engineering - Software Engineering for AI, CAIN, pp. 275–277. ACM (2024)
15. Rogers, E.M.: Diffusion of Innovations, 5th ed edn. Free Press, New York (2003)
16. Weske, M.: Business Process Management: Concepts, Languages, Architectures. Springer, Berlin (2024)

RCIS Doctoral Consortium

Digital Twins for Incident Detection and Response

Konstantinos E. Kampourakis(✉)

Department of Information Security and Communication Technology,
Norwegian University of Science and Technology, 2802 Gjøvik, Norway
konstantinos.kampourakis@ntnu.no

Abstract. Digital Twin (DT) technology is revolutionizing critical infrastructure (CI) sectors by enabling real-time monitoring, predictive analytics, and dynamic decision-making. However, this increased interconnectivity and complexity also introduces significant cybersecurity challenges. The current work investigates the potential of DTs to enhance cybersecurity incident detection and response mechanisms for CI systems. This investigation endeavors to explore how DTs, by leveraging machine learning and real-time data analysis, can identify cyber-specific anomalies, predict potential cyber threats, and simulate cyberattack scenarios to evaluate and optimize cybersecurity response strategies. Specifically, we emphasize the development of a comprehensive understanding of the role of DTs in enhancing system security and operational continuity, addressing evolving cyber threats with adaptive and proactive measures. Overall, our objective is to demonstrate the potential of DTs to build more resilient and secure infrastructures, safeguarding their reliability in a rapidly advancing digital landscape.

Keywords: Digital twin · critical infrastructure · incident detection · incident response

1 Introduction

The rapid advancement of Industry 4.0 and beyond technologies is revolutionizing the industrial manufacturing sector, giving rise to new information security challenges. Increasing connectivity, automation, and the use of smart systems, while driving efficiency and innovation, also create new threats, such as cyberattacks targeting control systems and the potential for cascading failures. In this context, the integration of Digital Twins (DTs) in critical infrastructure (CI) sectors, such as manufacturing, energy, and healthcare, is driving significant digital transformation. The National Institute of Standards and Technology (NIST) defines a DT as a "dynamic, real-time representation of physical objects, systems, or processes that mirrors the current and historical states of its physical counterpart, and leverages predictive capabilities for enhanced decision-making" [1]. By creating a virtual replica of the physical system, DTs can be used to simulate cyberattacks, identify vulnerabilities, and test the effectiveness of security measures before they are implemented in the real world. This ability to model and analyze complex systems in a virtual environment offers significant potential for enhancing cybersecurity within CI.

On the other hand, the expanded connectivity enabled by DTs in various CI sectors widens the potential attack surface for cybercriminals. As demonstrated by well-known incidents like Stuxnet [2], cyberattacks targeting Operational Technology (OT) infrastructures can have catastrophic consequences. To maintain operational integrity and prevent disruptions to essential services, it is imperative to implement robust cybersecurity measures to protect both the physical and digital components of Cyber-Physical Systems (CPS). In this context, DTs have evolved beyond just system simulation, now enabling real-time data-driven proactive incident detection and dynamic mitigation strategies as well. Nevertheless, as already mentioned, the increasing complexity and interconnectedness of CPS, powered by technologies like Internet of Things (IoT) and 5G and beyond, introduces new cybersecurity challenges. To ensure effective incident detection and response, it is crucial to address these challenges and implement secure communication protocols and robust data protection techniques to safeguard the transmission of data between DTs and physical assets. For instance, leveraging Machine Learning (ML) algorithms can enable DTs to analyze real-time data streams, identify anomalies, and predict potential cyberattacks with greater accuracy. However, research is still ongoing to develop and implement effective AI/ML-driven security solutions that can address the unique challenges posed by the dynamic and interconnected nature of DT-enabled CPS [3].

Contribution: This PhD thesis aims to explore how DT technologies can be leveraged to enhance cyber-incident detection and response in a range of CI sectors. The focus is on DT-based architectures capable of real-time and predictive modeling, as well as dynamic response to a range of cyberattacks, including malware, ransomware, Distributed Denial of Service (DDoS) attacks, insider threats, and zero-day exploits.

The remainder of this paper is structured as follows. The next section provides an overview of DTs in the context of CI and introduces their detection and response capabilities. Section 3 reviews the recent literature on DT-based incident detection and response, while Sect. 4 outlines key objectives and research questions. The research methodology and the expected results are detailed in Sect. 5. Section 6 concludes and provides future research directions.

2 Digital Twin Technology

Digital Twins (DTs) emerged in the 1960s at NASA as "living models" for the Apollo mission [4]. However, their widespread adoption accelerated after NIST's 2021 publication, "Considerations for Digital Twin Technology and Emerging Standards" [5], which clarified their definition, applications, and key considerations, driving academic and industrial interest. By leveraging DTs, production teams can analyze sensor data, machine data, and historical records to minimize defects, boost efficiency, and reduce downtime. In manufacturing, DTs shorten time-to-market by enabling virtual design and evaluation before production. However, the growing interconnectedness of DTs introduces significant cybersecurity risks, including malware attacks, data breaches, and manipulation. Robust security measures are essential to protect sensitive data and ensure DT integrity. This study explores how DTs enhance efficiency while mitigating cybersecurity risks. As mentioned in Sect. 1, this study focuses on DT applications in CI sectors, including energy, manufacturing, and healthcare. In the energy sector, DTs simulate cyberattack impacts on grid stability, helping predict failure scenarios and inform real-time decision-making. Security measures like encryption, access control, and IDS can safeguard DTs against cyber threats.

DT technology offers incident detection and response capabilities. As stated by NIST [6], a security incident is any event or series of events that compromise an information system's confidentiality, integrity, or availability. Such incidents, from unauthorized access to malware and phishing attacks, can lead to financial loss, reputational damage, and operational disruptions. Effective incident detection and response are crucial, and DTs, with their real-time simulation capabilities, can enhance security by identifying anomalies, predicting attack impacts, and testing response strategies. Unlike traditional security measures, DTs provide a dynamic, interactive environment for assessing and mitigating threats. The NIST Cybersecurity Framework (CSF) offers a flexible, risk-based approach to cybersecurity management. DTs can support its six core functions in incident detection and response.

- **Govern:** Defines risk management strategies, policies, and roles. DTs help simulate cybersecurity policies before implementation and model attack scenarios to assess mitigation measures.
- **Identify:** Involves asset identification, risk assessment, and vulnerability analysis. DTs provide a detailed inventory of system assets and simulate attacks to proactively identify weaknesses.
- **Protect:** Focuses on implementing security controls like firewalls, IDS, and encryption. DTs assess security controls in virtual environments and serve as training platforms, similar to cyber ranges and security testbeds [7,8], for cybersecurity procedures.
- **Detect:** Aims at identifying cyber threats through log analysis, IDS, and anomaly detection. DTs continuously compare real-world and virtual behaviors to detect anomalies and refine IDS strategies.
- **Respond:** Entails swift containment and mitigation of incidents. DTs simulate attacks to test response plans and aid forensic investigations by recreating attack scenarios.
- **Recover:** Restores operations post-incident through backup restoration and system re-imaging. DTs assist in testing recovery plans and restoring systems to a known good configuration.

Given the above, as discussed in Sect. 4, this work investigates the use of DT technology to enhance incident detection and response in CI sectors, aligning with the Govern, Identify, Detect, Respond, and Recover functions of the NIST cybersecurity framework. By creating virtual replicas of physical systems, DTs provide real-time insights, enabling proactive threat detection and early warning of cyberattacks through advanced analytics and ML. This study explores the technical challenges and opportunities of DT implementation in CI, focusing on security, resilience, and operational efficiency.

3 Related Work

This section provides a concise review of the most recent, relevant research published between 2020 and 2024 on the application of DT in various CI sectors such as energy, transportation, and healthcare. The related work discussed focuses on detection techniques to effectively identify potential threats and incident response strategies to minimize their impact. The search query used to identify the most relevant literature was "digital twins" **AND** ("cybersecurity" **OR** "cyber security" **OR** "cyber-security") **AND** ("incident detection" **OR** "incident response"). From these results, those directly related to CIs were selected. It is important to note that while the literature on incident detection and response is extensive, the application of DTs to support these security services in the context of CIs is still relatively nascent and largely unexplored."

Cybersecurity incident response is critical for mitigating the impact of cyberattacks on CI. Although traditional playbooks have been widely used, their effectiveness can be limited, especially in complex CPS. The authors in [9] introduce a novel approach to enhance cybersecurity playbooks using DTs. They provide a detailed analysis of how different DT modalities can be leveraged across various phases of incident response, and propose a systematic framework for optimizing playbook design. By applying their approach to a real-world scenario, they demonstrate its potential to improve the coordination of multidisciplinary teams and enhance overall resilience.

As previously mentioned, securing complex CPS is a critical challenge in today's interconnected and "cloudified" world. DTs offer a promising solution to enhance security by providing a virtual representation of the physical system. To this end, the authors in [10] introduce a novel framework that leverages DTs and advanced ML techniques to improve the security of CPS. Their framework incorporates state-of-the-art (SOTA) federated learning to enable advanced anomaly detection without compromising data privacy. Additionally, they address the security and trustworthiness of the DT itself by integrating blockchain technology to ensure secure communication and data storage. This novel approach has the potential to significantly enhance the resilience of CPS against cyberattacks.

Securing large-scale physical infrastructure, such as energy grids, is a key challenge. To address this hurdle, the authors in [11] demonstrate the application of DT technology to the energy sector, focusing on the security of sensor networks and Supervisory Control and Data Acquisition systems. By creating a digital replica of the physical infrastructure, they enable advanced simulation, anomaly detection, and incident response capabilities. This can help to mitigate the impact of cyberattacks and other threats, ensuring the continued operation of CI. DTs, as virtual replicas of physical systems, inherit the vulnerabilities of their real-world counterparts. This makes them attractive targets for cyberattacks, which can have severe consequences. In [12], the authors delve into the security challenges posed by DTs and propose a set of defensive strategies to mitigate these risks. They identify various types of attacks that can target DTs and discuss techniques to prevent and respond to such attacks, including intelligence-driven solutions, provenance-aware blockchain-based approaches, and fault-tolerant systems.

While DTs offer a promising avenue for enhancing the security and resilience of Industry 4.0 and beyond systems by enabling advanced simulation, testing, and monitoring capabilities, the increased interconnectivity between physical and virtual environments expands the attack surface, introducing new cybersecurity risks. In [13], the authors explore the cybersecurity implications of DTs in Industry 4.0, identifying potential vulnerabilities and threats. They highlight the importance of utilizing DTs as a critical component of a comprehensive security strategy, enabling the proactive identification and mitigation of risks. They conclude that by simulating cyberattacks and testing defensive strategies within the DT environment, organizations can strengthen their cybersecurity posture and reduce the likelihood of successful cyberattacks. Based on the above analysis and the study of the literature overall, below is a list of key challenges in the field of DTs for incident detection and response. These challenges emerge from both theoretical studies and practical implementations, reflecting the most pressing technical and operational obstacles faced by researchers and industry practitioners. While this list captures the primary concerns in the domain, the evolving nature of DT technology and cybersecurity necessitates continuous evaluation to address emerging risks and advancements.

- **Real-time synchronization:** Ensuring real-time synchronization between the physical and DTs is crucial for accurate incident detection and response. Challenges include data latency, network connectivity issues, and the complexity of modeling dynamic systems.
- **Data quality and quantity:** High-quality and sufficient data (benchmark datasets) is essential for training ML models and enabling accurate anomaly detection. Challenges include data sparsity, noise, and the difficulty of obtaining labeled data for specific incident scenarios.
- **Model accuracy and interpretability:** Developing accurate and interpretable ML models is critical for reliable incident detection. Challenges include model overfitting, underfitting, and the black-box idiosyncrasies of deep learning models [14].
- **Security and privacy:** Safeguarding the security and privacy of DTs is paramount, especially when dealing with sensitive data from CI. Challenges include data breaches, cyberattacks, and the risk of unauthorized access to critical information.

In addition to the challenges discussed above, several literature gaps can guide future research in this field.

- **_Standardized frameworks and methodologies:_** There is a need for standardized frameworks and methodologies for developing and deploying DTs for incident detection and response. This includes guidelines for data collection, model development, and validation.
- **_Multi-domain integration:_** As pinpointed previously, integrating DTs across different domains, such as energy, transportation, healthcare, and others, presents challenges. More research is needed to develop effective strategies for interoperability and data sharing.
- **_Human-in-the-loop:_** Incorporating human expertise into decision-making processes is crucial for effective incident response. More research is needed towards developing human-centered interfaces and collaborative tools for human-machine interaction.

4 Objectives and Research Questions

As previously noted, this PhD thesis aims to leverage DTs to improve cyberattack detection and response in CI sectors [15]. This overarching goal encompasses a number of key objectives. Given the increasing applicability of DTs across various CI sectors, we aim to explore how DT technology can enhance real-time detection, incident response, and recovery processes in CI networks. This will involve a comprehensive, systematic review of current methods and technologies that leverage DTs to counter cyberattacks, including their role in incident response, recovery, scenario simulation, and detection strategies. From a requirements engineering viewpoint, three key requirements are identified: First, the DT platform shall support real-time data ingestion and processing of CI network data to enable timely attack detection. Second, the platform shall support the simulation of diverse cyberattack scenarios against CI systems to facilitate incident response planning and training. Third, the platform shall integrate with existing CI cybersecurity tools and technologies (e.g., IDS, security information and event management systems) via APIs or standard protocols to enable coordinated incident response and recovery.

Building upon insights from the first key objective, this research will focus on designing a DT-based IDS specifically for CI networks. This will involve developing advanced ML algorithms that integrate with DT technology to identify anomalies and diagnose issues in these networks. The primary goal is to create adaptive models that can effectively detect cyber threats while minimizing false positives and negatives. An additional area of exploration will involve exploring how DT technology can be used to enhance cyber threat testing, including penetration testing, vulnerability scanning, red teaming, threat hunting, and attack surface analysis, to improve the efficiency of emergency response operations. This may include leveraging simulations and real-time analytics to predict potential incident impacts, optimize response strategies, and effectively allocate resources during critical events. Drawing upon the established objectives, the following research questions (RQs) are formulated to decompose each key objective into detailed, actionable inquiries and measurable goals.

- **RQ1**: What are the current SOTA approaches, challenges and limitations of using DTs for incident detection and response in CI cybersecurity? This involves four key tasks: firstly, to investigate how DTs can be utilized to accelerate incident response and recovery processes in CI networks; secondly, to explore current techniques for employing DTs to simulate diverse incident scenarios and evaluate the effectiveness, efficiency, and resilience of response strategies; thirdly, to identify the most promising and cutting-edge incident detection techniques leveraging DT technology, including anomaly detection, intrusion detection, and threat intelligence analysis; and finally, to conduct requirements engineering specifically for DT-based solutions in CI environments, as briefly described in the first key objective, encompassing performance, scalability, reliability, and security requirements.
- **RQ2**: What are the most effective ML and statistical techniques for detecting anomalies and diagnosing faults in CI networks using DTs, considering the potential for false positives and false negatives? This involves two key tasks: firstly, to develop adaptive ML-based intrusion detection algorithms, encompassing supervised, semi-supervised, and unsupervised approaches, with a strong emphasis on continuous learning capabilities to address the dynamic operational conditions and evolving threat landscape within CI environments; and secondly, to investigate how DTs can be used to proactively identify and mitigate cyber threats, including targeted attacks and data breaches, considering both external and internal sources of threats within CI networks.
- **RQ3**: How can DTs be used to enhance cyber threat testing, enabling more effective and proactive cybersecurity measures in CI networks? This involves three key tasks; firstly, to explore the development of DT-based systems to support various testing methodologies, including leveraging DTs for realistic attack simulations to enhance penetration testing, vulnerability scanning, red teaming, threat hunting, and attack surface analysis. Such simulations will provide a controlled environment to evaluate system performance under simulated attack scenarios, ultimately improving the efficiency of emergency response operations. Secondly, to investigate the use of DTs for real-time threat detection and response, encompassing the development of systems to monitor network traffic and identify anomalous behavior in real-time. This will be achieved by utilizing ML techniques to analyze large volumes of threat intelligence data and detect emerging threats, and exploring the automation of incident response procedures, including containment, eradication, and recovery, based on real-time threat intelligence derived from the DT environment. Thirdly, to enhance situational awareness and predictive analytics. This includes the utilization of DTs to create dynamic models of CI networks, enabling visualization and analysis of potential attack vectors, the development of predictive analytics models to forecast future cyber threats based on historical data and real-time intelligence, and the implementation of early warning systems to alert organizations of potential threats before they materialize.

5 Research Method and Expected Results

The objectives of this research proposal will be achieved through a structured, problem-centered approach, as outlined in Table 1. To ensure the practical applicability of the proposed architectures and systems, the well-known Design Science Research (DSR) methodology is followed [16]. With reference to Table 1, a systematic literature review will be conducted to identify gaps and challenges in the existing research and requirements engineering will be employed. Afterward, the focus will shift to threat modeling and the development of ML-based controls, accompanied by simulations to evaluate their efficacy. Addressing RQ3 will involve systematic literature review, cyber threat testing, and evaluation to validate proposed solutions and methodologies. This structured approach ensures comprehensive coverage of the research objectives and facilitates the development of robust, actionable outcomes.

Table 1. Methods for addressing RQs

RQ1	RQ2	RQ3
Systematic literature review	Threat modeling	Systematic literature review
Identify literature gaps and research challenges	Devise and assess ML-based controls	Cyber threat testing
Conduct requirements engineering	Simulation	Evaluation

Building upon this structured methodology, the next step is to translate these research methods into concrete outcomes that directly address the identified challenges. The expected results outlined below align with the objectives and research questions presented in Sect. 4. As already pointed out, to ensure the practical relevance of these results, the DSR methodology will be followed.

Identify the Problem: Conduct a comprehensive, systematic literature review on the use of DTs for incident detection and response in CI sectors. This review will identify limitations in existing IDS, such as their inability to adapt to dynamic conditions or address emerging threats. Additionally, the analysis will highlight gaps in incident response and emergency management, particularly in leveraging real-time data and simulations for informed decision-making.

Define Objects for the Solution: The planned research will articulate specific objectives for developing a DT-based solution to enhance cyber defense in CI networks. These objectives will include: (a) Developing adaptive ML-powered models for intrusion detection, (b) integrating DTs to improve situational awareness, and (c) optimizing emergency response strategies. The research will address challenges such as false alarms, evolving threat landscapes, and resource constraints.

Design and Development: A novel DT-enabled IDS will be designed and developed, incorporating SOTA ML algorithms for anomaly detection and fault diagnosis. The system will integrate predictive modeling and real-time analytics to simulate potential threats and optimize response actions. The development phase will ensure alignment with the previously defined objectives, incorporating insights from the problem identification stage.

Demonstration: The developed solution will be implemented and tested within a simulated CI network environment. This phase will showcase the system's ability to detect cyber threats, adapt to changes in operational conditions, and coordinate response efforts. Demonstration scenarios will mimic real-world incidents to illustrate the system's practical applicability and robustness.

Evaluation: The developed solution will undergo rigorous evaluation to assess its effectiveness in improving cyber defense for CI networks. Key performance metrics, including detection accuracy, adaptability to evolving threats, and response efficiency, will be analyzed. The system's performance will be validated through simulations and case studies to ensure it meets technical and operational requirements.

6 Conclusions

Integrating DTs into CI systems offers substantial gains in security, resilience, and operational efficiency. By creating real-time, dynamic virtual models, DTs facilitate proactive monitoring, predictive analysis, and automated responses. This is crucial for mitigating the escalating cybersecurity risks posed by the increasing complexity and interconnectedness of CI systems. This research emphasizes the crucial role of DTs in improving incident detection and response capabilities. By integrating ML and advanced data analytics, DTs facilitate accurate anomaly detection, threat prediction, and the simulation of attack and mitigation scenarios. This enhances both immediate incident response and long-term resilience against emerging cyber threats.

While offering significant advantages, DTs also present unique challenges related to secure communication, data integrity, and privacy. Addressing these, requires robust security frameworks, adaptive solutions, and interdisciplinary collaboration. In conclusion, DT technology has the potential to revolutionize CI cybersecurity, paving the way for more secure, sustainable, and resilient infrastructures in the digital age. Future research and implementation must prioritize addressing technical, ethical, and operational challenges to fully realize the potential of DTs in safeguarding critical systems.

Acknowledgements. This Ph.D. is carried out under the supervision of Prof. Vasileios Gkioulos of Norwegian University of Science and Technology, Gjøvik, Norway. This work has received funding from the Research Council of Norway through the SFI Norwegian Centre for Cybersecurity in Critical Sectors (NORCICS) project no. 310105.

References

1. NIST. Digital Twins for Advanced Manufacturing (2024). https://www.nist.gov/programs-projects/digital-twins-advanced-manufacturing. Accessed 16 Oct 2024
2. Farwell, J.P., Rohozinski, R.: Stuxnet and the future of cyber war. Survival **53**(1), 23–40 (2011)
3. Sarker, I.H., et al.: Explainable AI for cybersecurity automation, intelligence and trustworthiness in digital twin: methods, taxonomy, challenges and prospects. ICT Express (2024)
4. siemens. Apollo 13: The first digital twin (2024). https://blogs.sw.siemens.com/simcenter/apollo-13-the-first-digital-twin/. Accessed 16 Oct 2024
5. Voas, J., Mell, P., Piroumian, V.: Considerations for digital twin technology and emerging standards. Technical report, National Institute of Standards and Technology (2021)
6. Cichonski, P., et al.: Computer security incident handling guide. NIST Spec. Publ. **800**(61), 1–147 (2012)
7. Kampourakis, V., Gkioulos, V., Katsikas, S.: A stepby- step definition of a reference architecture for cyber ranges. J. Inf. Secur. Appl. **88**, 103917 (2025). https://doi.org/10.1016/j.jisa.2024.103917. ISSN: 2214-2126
8. Kampourakis, V., Gkioulos, V., Katsikas, S.: A systematic literature review on wireless security testbeds in the cyber-physical realm. Comput. Secur. **133**, 103383 (2023). https://doi.org/10.1016/j.cose.2023.103383. ISSN: 0167-4048
9. Allison, D., Smith, P., Mclaughlin, K.: Digital twin-enhanced incident response for cyber-physical systems. In: Proceedings of the 18th International Conference on Availability, Reliability and Security (2023)
10. De Benedictis, A., Esposito, C., Somma, A.: Toward the adoption of secure cyber digital twins to enhance cyber-physical systems security. In: International Conference on the Quality of Information and Communications Technology, pp. 307–321. Springer (2022)
11. Akerele, A., et al.: The digital twins incident response to improve the security of power system critical infrastructure. J. Comput. Sci. Coll. **39**(3), 86–99 (2023)
12. Suhail, S., Jurdak, R., Hussain, R.: Security attacks and solutions for digital twins. arXiv preprint: arXiv:2202.12501 (2022)
13. de Azambuja, A., et al.: Digital twins in industry 4.0-opportunities and challenges related to cyber security. Procedia CIRP **121**, 25–30 (2024)
14. Kampourakis, K.E., et al.: Balancing the act? Resampling versus imbalanced data for Wi-Fi IDS. Int. J. Inf. Secur. **24**(1), 47 (2025)
15. Lampropoulos, G., Larrucea, X., Colomo-Palacios, R.: Digital twins in critical infrastructure. Information **15**(8), 454 (2024)
16. Peffers, K., et al.: A design science research methodology for information systems research. J. Manag. Inf. Syst. **24**, 45–77 (2007)

Towards an Enterprise Architecture Based Approach for the Development of Digital Twins for Sustainable Real Estate Management

Marianne Schnellmann(✉)

Business Informatics, TU Wien, Vienna, Austria
marianne.schnellmann@tuwien.ac.at

Abstract. In the context of Real Estate Management, Digital Twins (DTs) can act as transformative tools for the integration of sustainability into construction practices by providing dynamic, data-driven virtual representations of physical assets. They enable stakeholders to optimise processes, reduce resource consumption, and align with sustainability objectives. However, effective implementation of DTs, in this context, remains limited by fragmented data ecosystems across different organisations, incomplete integration with regulatory frameworks, and insufficient consideration of sustainability goals. This is especially challenging, as access to high quality, and integrated, data is critical for DTs. Enterprise Architecture (EA) is generally used as an instrument to aid in managing and coordinating such challenges, as it provides a structured approach to addressing these challenges by aligning technical, organisational, and sustainability objectives. In line with this, we propose three primary research objectives: (1) assessing the current state of EA approaches for integrating DTs, (2) developing sustainability-oriented EA modelling concepts and design patterns, and (3) exploring the scalability of DTs from individual buildings to smart cities. Our aim is to position DTs as comprehensive decision-support tools for sustainability, enabling stakeholders to achieve better-informed decisions while addressing regulatory compliance, resource efficiency, and life cycle impacts.

Keywords: Digital Twin Engineering · Enterprise Architecture · Sustainable Real Estate Management

1 Introduction

The built environment sector is one of the most resource-intensive sectors globally; extending its influence far beyond physical structures. This sector is also responsible for at least 37% of global greenhouse gas emissions [27] and approximately 34% of global energy consumption [12,26]. The environmental impact spans the entire value chain, from material extraction and the production of construction components to the building and renovating structures. These figures underscore the sector's dual role as both a significant driver of climate change and a critical area for intervention [9]. Faced with these challenges, the transformation of the built environment to more sustainability in different aspects is not only necessary but also highly promising. In an era where societies and regulatory frameworks increasingly prioritise sustainability, initiatives such as the EU Taxonomy for sustainable activities and the European Green Deal have gained significant importance. These efforts aim to integrate environmental, social, and economic considerations into decision-making processes, ensuring long-term sustainable development. In addition to these comprehensive policies, numerous other measures, such as the official United Nations Sustainable Development Goals (SDGs), in particular, SDG 11: Sustainable Cities and Communities, contribute to advancing sustainability in real estate development.

The AECO (Architecture, Engineering, Construction & Operations) sector encompasses the entire built environment, managing all aspects from planning, design, construction, operation, maintenance, and end-of-life stages [6]. It plays a crucial role in driving sustainability by implementing energy-efficient designs, optimising resource use, and integrating innovative construction technologies, which results in making noteworthy contributions to the progress of the economy and society [29]. However, beyond the technical execution of sustainable buildings, there is a need to address their long-term economic viability, regulatory compliance, and strategic management. This is where Real Estate Management (REM) comes into play, leveraging these technological advancements to seamlessly integrate them into financial, regulatory, and strategic decision-making processes, with a strong emphasis on sustainability [5]. Building Information Modelling (BIM) [23] plays a central role in this transformation, serving as a digital planning method that consolidates all relevant building data into a unified model and enabling a comprehensive life cycle assessment (LCA). BIM primarily offers static data on the built environment and requires additional data sources to integrate real-time information into its models [25]. This is where the Digital Twin concept comes in; by dynamically linking with real-time data, the model is continuously updated and enables more precise analysis, monitoring and optimisation of the operation of buildings [6].

Digital Twins (DTs) are essential tools for enabling data-driven decision-making, particularly for optimising sustainability outcomes [14]. DTs create dynamic, virtual representations of physical assets by consolidating data from sensors, material databases, energy grids, and environmental sources, forming a foundation for LCAs, real-time monitoring, and predictive simulations. Data-driven DTs can only function accurately and effectively if the underlying data is trustworthy and coherent [2]. Ensuring the successful integration of DTs requires not only high-quality data but also a rigorous understanding of both creating and utilising such data to address fragmented data sources effectively [2]. DTs empower stakeholders to lower carbon footprints, minimise resource waste, and promote circular economy principles through strategic early-stage project interventions [1]. Integration with BIM further enhances these capabilities, enabling energy efficiency simulations and life cycle optimisations [28]. By addressing the relationship between data integration and DT functionality while ensuring data integrity, stakeholders can maximise the value of DTs as decision-support tools. This approach transforms sustainability decision-making by aligning environmental benefits, cost efficiency, and regulatory compliance, revolutionising industry practices and driving long-term improvements.

Despite their potential, DTs face persistent challenges in capturing the full complexity of sustainability objectives. While DTs excel in providing robust visualisations and data representations, their effective application for informed, dynamic, life cycle-wide decision-making on different scales remains a major hurdle [20]. The effectiveness of DTs is limited by fragmented data ecosystems and insufficient integration, which constrain their ability to comprehensively address goals such as those outlined in global climate agreements. Overcoming these challenges is crucial to unlocking the transformative potential of DTs as tools for driving sustainable innovation and decision-making.

Enterprise Architecture (EA) is generally positioned as an instrument that can (amongst others) aid in managing and coordinating such challenges, as it provides a structured approach to addressing these challenges by aligning technical, organisational, and sustainability objectives. In line with this, the goal of the planned doctoral research project is to establish an EA-based approach for the development of DTs to support sustainable REM.

The aim of this paper, which is positioned at the early stages of doctoral research, is to gather additional insights, refine existing ideas, and identify potential research directions by engaging with peers and collecting diverse perspectives by participating in the Doctoral Consortium. By focusing on understanding the role of DTs in advancing sustainability within the AECO sector and REM in particular, the planned research project seeks to contribute to the broader discussion on sustainability challenges and explore innovative pathways for integrating DTs into decision-making processes for a more sustainable built environment.

The remainder of this paper is structured as follows. In Sect. 2, we discuss the problem formulation. Following that, the research design in Sect. 3 elaborates on objectives, questions, and methodologies, leading to the evaluation discussion in Sect. 4, including validation techniques and highlighting this thesis's contribution. Finally, in Sect. 5, we draw the paper to a conclusion.

2 Problem Formulation

DTs carry the promise of being transformative tools, offering dynamic, data-driven virtual representations of physical assets that enable stakeholders to optimise processes, reduce resource consumption, and achieve sustainability objectives. However, the effective implementation of DTs in sustainability-focused applications remains limited by fragmented data ecosystems, siloed processes, and insufficient integration with regulatory and organisational frameworks [13,22].

One of the primary hurdles is the lack of high-quality, integrated data necessary for effective DT functionality. While Building Information Modelling (BIM) and material databases[1] such as ÖKOBAUDAT, baubook, and Dataholz provide essential inputs these resources often suffer from incompleteness or inconsistencies. This limits the ability to effectively create comprehensive LCAs or implement circular economy principles. Additionally, the integration of sustainability metrics, such as carbon footprint and resource efficiency, with existing DT frameworks remains fragmented, often addressing isolated aspects of projects rather than providing a holistic perspective [13,16].

A practical example that illustrates these challenges is the sustainable renovation of existing buildings. In the later stages of the operational phase, decision-makers – such as real estate trustees – are responsible for evaluating whether to renovate, partially or fully dismantle existing buildings, or implement alternative measures to improve the building performance. These decisions are often hindered by a lack of integrated, structured information [15], while trustees also face increasing pressure to balance sustainability goals with economic constraints [21]. More broadly, building renovation is characterised by a lack of consensus on key criteria, methodologies, and decision-support tools, which frequently hinders the implementation of the most sustainable and effective solutions [15]. Although Building Twins evolve from the BIM and exist as the As-Built-Models after the construction phase [11], they often lack sustainability data and the capability to perform scenario-based analyses [19]. This significantly limits their potential to support renovation strategies that aim to balance environmental performance, economic viability, and legal compliance. Furthermore, the absence of a common set of decision criteria or structured support tools results in ad-hoc decision-making processes that often fall short of achieving long-term sustainability goals.

EA provides a structured framework to address such challenges by aligning technical, organisational, and sustainability objectives throughout project life cycles. An agile and resilient architectural approach enhances responsiveness to the evolving demands of sustainability [24]. EA-based frameworks have been shown to enhance data integration, operational efficiency, and decision-making in complex systems. For example, Wu et al. [30] demonstrated the potential of a DT-based EA control platform in managing complex infrastructures, while Atencio et al. [4] explored the integration of BIM and EA frameworks to streamline planning in construction megaprojects. Building on these principles, Lick et al. [17,18] developed practitioner-driven approaches using capability maps to support digital transformation and decision-making, and Antunes et al. [3] introduced a reference architecture utilising ArchiMate to bridge theoretical and practical applications of DTs. Furthermore, Edrisi et al. [7,8] proposed a comprehensive EA blueprint, emphasising organisational resilience and sustainability. These related works highlight the transformative potential of integrating EA principles into DT frameworks, emphasising their scalability and alignment with long-term sustainability goals.

This research builds on these insights by proposing an EA-inspired framework that addresses the challenges of data fragmentation, scalability, and sustainabil-

---

[1] https://www.oekobaudat.de/, https://www.baubook.at, https://www.dataholz.eu/.

ity integration. As such, the proposed approach aims to enhance the role of DTs as comprehensive decision-support tools, enabling more effective sustainability monitoring and compliance with global climate objectives.

3 Research Design

This section explains the central research objectives and research questions that form the core of the study, as well as the methodological approach used to answer the research questions and achieve the defined objectives.

3.1 Research Objectives and Questions

The research objectives are structured into three parts, each addressing a specific dimension of the research questions.

I. Status Quo The first objective is to establish a comprehensive foundation for understanding how EA approaches can be applied (initially in a sector-agnostic manner) to integrate and model DTs within an organisation's architecture. This includes identifying and analysing holistic EA frameworks and methodologies that go beyond treating DTs as isolated systems for specific purposes, instead of situating them as integral components within the larger organisational context. Therefore, we want to answer the research questions: (RQ1) *Which EA-related approaches exist in the literature for integrating DTs?* and (RQ2) *What are the limitations of existing EA approaches in supporting the seamless integration of DTs?* Building on this foundation, the research focuses explicitly on the role of DTs in promoting sustainability within the AECO sector. The study investigates how sustainability requirements and regulations are incorporated into DTs and the extent to which detailed sustainability-related information is accessible through these systems and answers the research questions: (RQ3) *What sustainability information is provided in the AECO sector's DTs, and to what level of detail is it available?*, (RQ4) *In which life cycle phases of a building is sustainability information provided and which data is used in the respective phases?* and (RQ5) *How detailed is the sustainability information that a DT receives and that it represents?*

II. EA-based modelling concepts & design patterns As a next objective, we aim to investigate the extent to which existing EA modelling approaches can accommodate the design of DTs and their sustainability functions. We will analyse whether the current approaches can be extended or optimised or whether new modelling concepts and/or design patterns are required to meet the specific requirements of sustainability and digital transformation. One focus is on exploring the role of DTs as a basis for decision-making by utilising embedded sustainability information that is aligned with company-specific priorities and measures. The research examines how DTs can embed sustainability principles such as circular economy, LCA, and regulatory compliance into both strategic and operational decision-making.

In addition, the research will analyse how DTs facilitate the integration of sustainability criteria into operational and strategic frameworks. A particular focus is on the extent to which DTs can provide actionable insights to harmonise sustainability goals with long-term business strategies. The aim is to assess the transformative potential of DTs and explore their contribution to promoting sustainable practices in the AECO sector. This research will ultimately develop approaches for design patterns that can model and operationalise DTs both at the level of individual organisations and at an overarching level – for example, in relation to sustainability aspects. With this, we want to answer (RQ6) *How effectively do current EA approaches support the integration of DTs into company systems?*, (RQ7) *To what extent do existing EA frameworks need to be extended or adapted to fulfil the specific requirements of DT implementations?* and (RQ8) *How can new or existing design patterns promote interoperability, scalability and efficiency in DT integrations?*

III. Digital Twin in-the-small – in-the-large scale The research objective is to explore how DTs can be effectively scaled to support comprehensive sustainability efforts within the AECO sector. In particular, the increasing complexity and the changing requirements and functionalities that arise when the scope is expanded from individual buildings to neighbourhoods and whole (smart) cities are considered. It also examines how the sustainability factors develop due to the different scopes and what changes occur in the priorities. In addition to these aspects, the research addresses the architectural implications of scaling, including the need for flexible and modular EA structures that can accommodate distributed systems and cross-organisational coordination. Furthermore, data governance challenges – such as ensuring interoperability, managing data responsibilities, and maintaining transparency across stakeholders – are considered critical elements to enable reliable and scalable DT ecosystems. In this regard, we want to answer the research questions: (RQ9) *How does EA contribute to enhancing the scalability and strategic alignment of DTs across small- and large-scale organisational ecosystems?* and (RQ10) *How can EA frameworks incorporate DTs for predictive analytics and operational efficiency?*

3.2 Research Methodologies

This doctoral research adopts a Design Science Research (DSR) [10] approach to investigate and address the research objectives systematically. The methodology is structured into three parts, aligned with the research objectives: Understanding the status quo, developing EA-based design patterns (artefact), and a method for scaling DT implementations. A variety of complementary methods are employed for each part to ensure comprehensive insights and practical relevance.

The **I** part focuses on understanding the status quo by conducting a systematic literature review (SLR) to analyse existing EA frameworks and their applications in integrating DTs. This includes exploring both domain-unspecific and AECO-specific contexts. A detailed examination of the representation of sustainability information, such as life cycle assessment (LCA) data, within DTs in the AECO sector will identify current gaps and inconsistencies. Additionally, a gap analysis will be performed to compare the existing capabilities of EA frameworks and DT implementations against the sustainability requirements of the AECO sector. To supplement these findings, expert interviews will be conducted with professionals in EA and DTs, focusing on sustainability, to provide qualitative insights into the limitations and potential of DT-EA integration. Case studies from real-world applications will further illustrate practical challenges, the utilisation of data, and the sustainability impact of DT implementations.

The **II** part focuses on developing EA-based design patterns to address identified gaps and extend the capabilities of existing frameworks. Conceptual modelling techniques will be used to design or adapt EA frameworks that integrate DTs, incorporating elements such as circular economy principles, LCA, and scenario simulations. These models will then be refined through workshops and co-creation sessions with stakeholders, including architects, engineers, and sustainability experts, to ensure relevance and applicability. To evaluate the practical utility of the design patterns, simulation tools will be employed to test them in various AECO scenarios, focusing on interoperability, scalability, and sustainability. Prototypes of the design patterns will be developed, and iterative testing and feedback processes will guide further refinements.

The **III** part explores scaling DT implementations to enable effective sustainability monitoring in the AECO sector. Comparative analyses of DT implementations at different scales – ranging from individual buildings to neighbourhoods and smart cities – will identify scalability challenges and opportunities. Field studies, including pilot projects in varying contexts, will assess how scaling influences sustainability management and monitoring. Multi-stakeholder collaborations will be facilitated to address inter-organisational data integration and governance challenges. To capture the long-term evolution of DTs, longitudinal studies will observe changes in sustainability priorities as implementations scale. Additionally, design thinking workshops will involve diverse stakeholders in developing innovative solutions for scaling DTs effectively while maintaining a strong focus on sustainability goals.

4 Evaluation Discussion

Validation Techniques

A combination of comprehensive validation techniques will be employed to ensure the robustness and applicability of the proposed approach. These techniques aim to assess both the theoretical soundness and the practical feasibility and impact of the developed methods and artefacts in real-world settings.

A key validation method is expert validation, involving professionals in the fields of EA, DTs, and sustainable built environments. These experts will evaluate critical aspects such as integrating sustainability features, scalability, and alignment with industry standards. Through structured interviews and workshops, actionable feedback will be collected to identify potential gaps, refine the artefacts, and ensure their practical relevance. Complementing this, field experiments will be conducted to test the artefacts in real-world environments, e.g., within a construction project. Observing how practitioners interact with the developed solutions in their workflows will provide valuable insights into usability and integration into existing systems.

To evaluate the effectiveness of the proposed model in a structured and transparent manner, several key dimensions will be considered. These include the perceived usability and user acceptance of the artefacts, assessed through usability tests. Furthermore, the impact on decision-making quality will be explored through comparative scenario analyses, examining the extent to which the artefacts support more informed and sustainable decisions. In addition, the scalability of the developed concepts will be evaluated by applying them across different implementation contexts, highlighting how sustainability priorities evolve with scale.

This multidimensional evaluation ensures that both the conceptual design and practical application of the artefacts meet the needs of stakeholders and are aligned with the overarching research goals.

Contribution of the Thesis

This thesis advances the integration of DTs into EA by developing scalable and adaptable design patterns that align technical, organisational, and sustainability objectives. It addresses key challenges in interoperability, data management, and sustainability, enabling DTs to support actionable insights, scenario simulations, and strategic decision-making across organisational ecosystems, from individual projects to smart cities.

The primary audience for the research results includes researchers and practitioners in DT engineering and EA implementation, particularly those seeking scalability, interoperability, and sustainability solutions. The findings also benefit organisations aiming to improve cross-disciplinary collaboration, optimise processes, and reduce resource waste.

The thesis adheres to open science principles by making all results and artefacts openly available on GitHub and publishing research papers as open access. This ensures transparency, fosters collaboration, and supports further progress in business informatics.

5 Conclusion

This research lays the groundwork for developing an EA-based approach to integrating DTs for sustainable REM. The next steps include two systematic literature reviews: existing EA approaches for DT integration and sustainability data in DTs within the AEC sector. These reviews, combined with insights from this consortium, will refine the research direction and ensure practical relevance, advancing the alignment of DTs with sustainability goals and scalable EA frameworks.

Supervision

This research work has been conducted under the supervision of Prof. Henderik Proper at the TU Wien, and has not been presented at any conference or doctoral symposium yet.

References

1. Adams, K.T., Osmani, M., Thorpe, T., Thornback, J.: Circular economy in construction: current awareness, challenges and enablers. Proc. Inst. Civil Eng. - Waste Resour. Managem. **170**(1), 15–24 (2017)
2. Afzal, M., et al.: Delving into the digital twin developments and applications in the construction industry: a PRISMA approach. Sustainability **15**(23) (2023). https://doi.org/10.3390/su152316436
3. Antunes, J., Barata, J., da Cunha, P.R., Estima, J., Tavares, J.: A reference architecture for dry port digital twins: preliminary assessment using ArchiMate. Lect. Notes Bus. Inf. Process. **513**, 131–145 (2024)
4. Atencio, E., Rivera, F.M.L., Mancini, M., Bustos, G.: Towards the integration between construction projects and the organization: the connections between BIM and enterprise architecture. In: Favari, E., Cantoni, F. (eds.) Complexity and Sustainability in Megaprojects, pp. 161–176. Springer Nature Switzerland, Cham (2023). https://doi.org/10.1007/978-3-031-30879-6_13
5. Christensen, P.: Key strategies of sustainable real estate decision-making in the united states: a Delphi study of the stakeholders. Ph.D. thesis, Clemson University (2012)
6. Deng, M., Menassa, C.C., Kamat, V.R.: From BIM to digital twins: a systematic review of the evolution of intelligent building representations in the AEC-FM industry. J. Inf. Technol. Constr. **26** (2021)
7. Edrisi, F., Perez-Palacin, D., Caporuscio, M., Giussani, S.: Developing and evolving a digital twin of the organization. IEEE Access **12**, 45475–45494 (2024). https://doi.org/10.1109/ACCESS.2024.3381778
8. Edrisi, F., et al.: EA blueprint: an architectural pattern for resilient digital twin of the organization. Commun. Comput. Inf. Sci. **1462**, 120–131 (2021). https://doi.org/10.1007/978-3-030-86507-8_12
9. European Commission: Buildings and construction (nd). https://single-market-economy.ec.europa.eu/industry/sustainability/buildings-and-construction_en. Accessed 24 Jan 2025
10. Hevner, A.R., March, S.T., Park, J., Ram, S.: Design science in information systems research. MIS Q., 75–105 (2004)
11. IG Lebenszyklus Bau: Leistungsbilder in den Projektphasen (2017) (2024). https://ig-lebenszyklus.at/wp-content/uploads/2024/06/Innenteil_01.pdf. Accessed 24 Feb 2025
12. International Energy Agency: Global CO2 emissions from buildings, including embodied emissions from new construction, 2022 (2023). https://www.iea.org/data-and-statistics/charts/global-co2-emissions-from-buildings-including-embodied-emissions-from-new-construction-2022. Accessed 14 Jan 2025
13. Jahangir, M.F., Schultz, C., Kamari, A.: A review of drivers and barriers of digital twin adoption in building project development processes. J. Inf. Technol. Constr. **29**, 141–178 (2024)

14. Jeong, Y., Flores-García, E., Wiktorsson, M.: A design of digital twins for supporting decision-making in production logistics. In: 2020 Winter Simulation Conference (WSC), pp. 2683–2694 (2020). https://doi.org/10.1109/WSC48552.2020.9383863
15. Kamari, A., Corrao, R., Kirkegaard, P.H.: Sustainability focused decision-making in building renovation. Int. J. Sustain. Built Environ. **6**(2), 330–350 (2017). https://doi.org/10.1016/j.ijsbe.2017.05.001
16. Khodeir, L.M., Othman, R.: Examining the interaction between lean and sustainability principles in the management process of AEC industry. Ain Shams Eng. J. **9**(4), 1627–1634 (2018). https://doi.org/10.1016/j.asej.2016.12.005
17. Lick, J., et al.: Digital factory twin: a practioner-driven approach for integrated planning of the enterprise architecture **128**, 603–608 (2024)
18. Lick, J., Weller, J., Brock, J., Pathak, S., Disselkamp, J.P., Kühn, A., Dumitrescu, R.: Guiding the transformation to a digital factory twin: towards an enterprise-architecture-management-based approach with the help of a capability map **130**, 736–742 (2024)
19. Lõhmus, D., Pikas, E.: Creation of digital twin models for renovation: an integrative literature review. In: Proceedings of the 2023 European Conference on Computing in Construction and the 40th International CIB W78 Conference. EC3 2023, European Council for Computing in Construction (2023). https://doi.org/10.35490/ec3.2023.307
20. Najafi, P., Soltani, A., Ahmad Khan, A., Chizfahm, M., Sepasgozar, S., Gu, N.: Digital twin for urban decision support systems: scientometric and thematic analysis, pp. 110–145. Routledge Taylor & Francis Group, United Kingdom (2024)
21. Nyoni, V., Broberg Piller, W., Vigren, O.: Sustainability action in the real estate sector - an organizational and institutional perspective. Cleaner Prod. Lett. **5**, 100049 (2023). https://doi.org/10.1016/j.clpl.2023.100049
22. Rafsanjani, H.N., Nabizadeh, A.H.: Towards digital architecture, engineering, and construction (AEC) industry through virtual design and construction (vdc) and digital twin. Energy Built Environ. **4**(2), 169–178 (2023). https://doi.org/10.1016/j.enbenv.2021.10.004
23. Sacks, R., Eastman, C., Lee, G., Teicholz, P.: BIM Handbook: A Guide to Building Information Modeling for Owners, Designers, Engineers, Contractors, and Facility Managers. John Wiley & Sons (2018)
24. van Schalkwyk, P., Isaacs, D.: Achieving Scale Through Composable and Lean Digital Twins, pp. 153–180. Springer International Publishing, Cham (2023)
25. Tang, S., Shelden, D.R., Eastman, C.M., Pishdad-Bozorgi, P., Gao, X.: A review of building information modeling (BIM) and the internet of things (IoT) devices integration: Present status and future trends. Autom. Constr. **101**, 127–139 (2019). https://doi.org/10.1016/j.autcon.2019.01.020
26. United Nations Environment Programme: Global Status Report for Buildings and Construction: Beyond foundations: Mainstreaming sustainable solutions to cut emissions from the buildings sector (2024). https://doi.org/10.59117/20.500.11822/45095
27. United Nations Environment Programme, Yale Center for Ecosystems + Architecture: Building Materials and the Climate: Constructing a New Future (2023). https://wedocs.unep.org/20.500.11822/43293. Accessed 14 Jan 2025
28. Walasek, D., Barszcz, A.: Analysis of the adoption rate of building information modeling [BIM] and its return on investment [ROI]. Procedia Eng. **172**, 1227–1234 (2017). https://doi.org/10.1016/j.proeng.2017.02.144

29. Waqar, A., et al.: Examining the impact of BIM implementation on external environment of AEC industry: a PEST analysis perspective. Dev. Built Environ. **17**, 100347 (2024). https://doi.org/10.1016/j.dibe.2024.100347
30. Wu, J., et al.: Research and design of a digital twin-based enterprise architecture digital control platform for provincial electrical power company, pp. 186–196 (2021)

A Method for Domain Reference Model Inference Through Knowledge and Data Intelligent Unifiers

Pedro Guimarães[1,2]

[1] ALGORITMI Research Centre, University of Minho, Campus de Azurém,
4800-058 Guimarães, Portugal
id11164@alunos.uminho.pt
[2] CCG/ZGDV Institute, University of Minho, Campus de Azurém,
4800-058 Guimarães, Portugal

Abstract. Transforming and integrating heterogeneous datasets into structured and semantically enriched data models remains a critical challenge in data and knowledge engineering. Addressing this challenge requires a systematic approach, as data and knowledge integration is often the most time-consuming phase. This work proposes a model to identify the key dimensions for establishing a method to infer Domain Reference Models. Furthermore, this work also proposes how such method can be decomposed into components and steps that encompass data normalization, advanced knowledge and data alignment, and graph-based representations, in order to infer and integrate Domain Reference Models.

Keywords: Domain Reference Model · Knowledge · Data · Knowledge Graphs

1 Introduction

The process of transforming and integrating heterogeneous datasets into structured and semantically enriched models remains a critical challenge in data and knowledge engineering. This task is compounded by the variability in data formats, lack of semantic consistency, and the need to align data and knowledge with established standards and data models. Moreover, the absence of systematic methods, processes and tools often results in inefficient processing pipelines, limiting the ability to integrate, analyze, and extract value from diverse datasets. In addition, experts may model the same domain differently, reflecting their unique perspectives and understanding, and such models may even vary with time.

Reference models, often referred to as "universal models, generic models, or model patterns" [2], are crucial for understanding domain-specific structures. [3] expanded this concept by introducing the importance of the expert and their domain knowledge, defining a Reference Model as a domain-specific structure that clearly expresses the expertise produced by an expert or body of experts. In the context of this work, a Domain Reference Model (DRM) should provide domain-specific knowledge through a graph-structured data representation, such as Knowledge Graphs (KGs). Additionally, it should support partial views of a specific domain or its subdomains and provide the means to ensure it remains updated through continuous knowledge and data integration. This is particularly important for organizations that are structured into complex multiple domains or subdomains, with a focus on integrating new business processes and datasets, providing an integrated view of organizational data [14]. To address the aforementioned challenges, we propose a model designed to identify the dimensions that need to be considered by a method that aims to infer and evolve DRMs. Moreover, we also propose how such method can decompose the established dimensions into a series of steps that encompass data normalization, advanced knowledge and data alignment, and graph-based representations, in order to infer and integrate DRMs.

The remainder of this paper is organized as follows. Section 2 reviews relevant literature on using KGs to support data and knowledge integration. Section 3 outlines the research objectives and questions guiding the study. Section 4 details the methodology employed. Section 5 introduces the model and describes its dimensions. Then, the proposed method is detailed. Finally, the work concludes with Sect. 6 summarizing the current work and future directions.

2 Background

According to the definition by [10], a reference model provides understanding of the meaningful relationships between entities within a given domain. Additionally, it supports the development of specific references or concrete architectures by applying consistent standards or supporting specifications for the domain. [1] categorizes reference models into different variants, one of which is the "Domain-Specific Models". These models are developed within the context of a sector's specific functions and aligned with its inherent processes. [13] also emphasizes that models typically embody best practices for a specific sector, offering a generic solution that lowers the costs of model design while simplifying organizational management and control.

These models encapsulate domain knowledge and support the development of domain specific processes [12]. The authors conclude that using reference models require guidance in their reuse and flexibility in adapting them to the unique requirements. However, a significant challenge in updating a model is the requirement to reevaluate the knowledge and data each time new information is added [17]. This is due to new knowledge and data that can potentially change the context or validity of previous integrations (become outdated [7,17]) in later integrations. As a result, not only must the new knowledge and data be integrated, but the entire model must also be reassessed to ensure that all integrations remain coherent and accurate.

KGs are data models that can represent and extract knowledge using deductive and inductive techniques, integrating information from diverse, dynamic, and large-scale collections of data [6,9,15]. KGs serve as a versatile framework for representing such knowledge, relying on the extraction and analysis of entities and their relationships [19]. Due to their flexible structure, KGs offer a promising approach to this challenge. However, traditional KGs do not incorporate evolving features, where knowledge dynamically grows and changes over time to include continuously emerging new facts [8]. In the process of incremental construction of KGs, the input encompasses not only the new data intended for addition, but also the existing version of the KG [5]. The authors argue that evolving KGs depict interactions across different moments, adding a temporal dimension that traditional KGs cannot model due to their inherent incompleteness.

3 Research Objectives and Questions

In this section, we outline the main goals of our research and the specific questions we aim to answer. The objectives represent the broader goals we hope to achieve, while the questions break down goals into smaller, more focused parts. The following subsections explain each objective and question in more detail.

3.1 Research Objectives

This work aims to develop a method for automated inference and evolution of DRMs. The objectives outlined below aim to address the challenges of knowledge and data integration towards unified and evolving DRMs:

- **RO1:** Propose a model that encompasses the dimensions and components that must be considered in order to establish a method that allows one to infer DRMs.
- **RO2:** Propose a method that involves multiple components (identified in RO1) and a series of steps supported by techniques or tools for knowledge and data integration of multiple sources to infer DRMs.
- **RO3:** Benchmark methods (such as algorithms and tools) that enhance the evolution of a DRM. To do so, we will set controlled experiments using the same input data and compare how effectively the methods address DRM evolution. In this process, we will explore, review, and define the appropriate metrics to conduct a systematic assessment.
- **RO4:** Assess the proposed model (RO1) and method (RO2) with an instantiation in obtaining an evolving and integrated DRM that accurately reflects its domain. At the end of each DRM iteration, we will validate the outcomes with domain experts through controlled experiments using real industrial data.

3.2 Research Questions

The primary goal of this step is to review existing knowledge and data integration studies, addressing the following challenges: (1) its evolving nature and ability to grow and integrate new knowledge; (2) handling obsolete or outdated knowledge and facts; and, (3) use of methods to perform the integration of new knowledge or data. Despite acknowledging that alternative methodologies and data representations exist, in our work, we will instantiate our method using KGs as the primary approach. The characteristics of KG in representing data and knowledge, the integration with advanced methods, such as AI, and the results of the background review shows potential in following this approach. Based on this, the following research questions (RQs) were defined:

- **RQ1:** What are the knowledge and data integration challenges towards a domain reference model using KGs?
  As referred in the survey conducted by [16], to achieve a unified domain model (using KGs), domain experts should address challenges such as: maintenance (evolution), automated knowledge extraction from graphs, and understanding the human role. In the same way, this question seeks to identify the current challenges and limitations of knowledge and data integration towards a DRM using KGs, providing support for the RO1 and RO2.

- **RQ2:** What artifacts (methods, models, approaches or frameworks) are being used to address the knowledge and data integration challenges in KGs?
  To tackle current challenges in knowledge and data integration, various techniques have been identified, as highlighted in the study by [18]. This research question seeks to explore the most commonly employed methods, models, tools, and frameworks for addressing integration challenges. Additionally, it aims to determine whether these solutions are tailored to specific challenges or applied in a broader, more generalized context. The findings will directly contribute to RO3.

- **RQ3:** How are existing approaches handling the evolving nature of KGs, including new and outdated knowledge and data?
  The research of [8] identifies key differences between traditional and evolving KGs. Thus, traditional KGs do not incorporate evolving features, where knowledge dynamically changes over time to include continuously emerging new facts. The authors argue that evolving KGs depict interactions across different generation times, adding a temporal dimension that traditional KGs cannot model, due to their inherent incompleteness. This research question aims to evaluate the extent to which existing approaches address the evolving nature of KGs. It also investigates whether current methods account for assessing knowledge and facts following the integration of new information. The findings will directly contribute to achieving RO1 and RO2.

4 Research Design

This research follows the Design Science Research Methodology (DSRM) for Information Systems [4]. Although these activities are expected to occur sequentially, the methodology allows for initiation at any activity and returning to previous activities whenever necessary [11]. It emphasizes iterative development and refinement, a core principle of DSRM, through the planned iterations of knowledge and data integration. The approach is problem oriented, focusing on the real-world challenge of creating an evolving DRM. The DSRM allows for four possible entry points to the research process. Below, this is discussed in more detail:

- **Problem Identification and Motivation:** The motivation for combining multiple data sources is concerned with the need to provide a unified view that facilitates the integration of heterogeneous data into a common structure. This step includes a literature review to understand the gaps of current knowledge and data integration approaches towards a DRM.
- **Definition of the Objectives of a Solution:** This step involves defining clear objectives to guide the research. Specifically, it focuses on identifying and characterizing the dimensions and components that a DRM must consider, as well as the methods required for iterative and continuous knowledge and data integration. To accomplish this, interviews with domain experts will be conducted. Building on this, the objectives that comprise our main research goal, consist of benchmarking artifacts (assessed in RO3) that support the continuous integration; Instantiate and assess the proposed method to revise the aforementioned integration and DRM evolution.
- **Design and Development:** We start by designing the model, detailing dimensions and components essential for DRM evolution, mainly for knowledge and data integration, such as data normalization and data versioning, or, knowledge and data extraction. Then, we develop the method that associate the aforementioned components to specific steps to integrate knowledge and data from multiple source models into a DRM.
- **Demonstration:** We begin by implementing the method in controlled experiments designed to simulate common integration challenges, showcasing its ability to unify diverse knowledge and data perspectives within the same domain. Subsequently, the proposal will be applied to an industrial case study, where it will be evaluated through a real-world integration scenario.
- **Evaluation:** We assess the effectiveness and efficiency of the proposed model, method and of the obtained DRM in reflecting the domain evolution. The evaluation criteria will be based on the objectives outlined and qualitative feedback from domain experts through the aforementioned interviews.
- **Communication:** We document the aspects of our research, from design, development, and evaluation results. The research findings and implications will be documented and shared with both academic and practitioner communities interested in DRMs, and data and knowledge integration. This includes publishing in journals, conferences, and creating open-source code or frameworks for community use.

5 Proposed Approach

From understanding domain concepts to achieving domain specific models, the proposed approach addresses how to drive a seamless progression from datasets to domain-specific models by integrating advanced normalization, map with existing knowledge, and assessment pipelines. The model may leverage technological artifacts (such as, Large Language Models (LLMs) that will be evaluated in RO3) to support domain knowledge extraction, completion and alignment processes, reducing manual effort and potentially enhancing the quality and consistency of extracted data and knowledge. Also, by retaining a graph-based perspective within the DRM representation, the resulting partial or full views support higher-level analytical tasks, including reasoning and predictive modeling. The next sections present the current work that has been done to address the ROs. Thus, the following sections are divided in one RO per subsection.

5.1 RO1: DRM Main Dimensions and Components

The proposed model (aligned with the RO1) aims to identify the dimensions and components that need to be considered by a method that allows one to infer DRMs. The model is conceptualized and presented (see Fig. 1) using a simplified version of the Unified Modeling Language (UML) Class Diagram notation. The dimensions are depicted as rectangles with a grey background. The main components are depicted as rectangles, omitting attributes to enhance the model's readability. The structure of dimensions and components was intentionally designed anticipating the need to decompose them further as the work evolves.

The **Domain Knowledge** dimension encapsulates expertise and structured rules relevant to the domain. The **Data Normalization** dimension deals with reconciling and standardizing data to maintain consistency and interoperability. The Data Pipeline orchestrates the core concepts: Data Collection, Data Merging, Data Transformation, Data Mapping, and Data & Knowledge Extraction, ensuring data is systematically gathered, harmonized, and prepared for further use. The **Data Representation** dimension produces partial or full views of the DRM tailored to business or technological requirements. Lastly, the Data and Knowledge component serves as a central component in the model, interacting with all other components.

5.2 RO2: Method for Inferring and Evolving DRMs

The proposed method, depicted in Fig. 2, is organized into interconnected dimensions and components (from RO1, Fig. 1), with the additional detail that is required in order to establish a series of steps that correspond to stages of knowledge and data integration. This way, the components operate sequentially, leveraging a variety of supporting methods and tools to drive the inference, representation, and evolution of the DRM. The method addresses two main challenges: (1) Transform heterogeneous, semi-structured data into semantically data

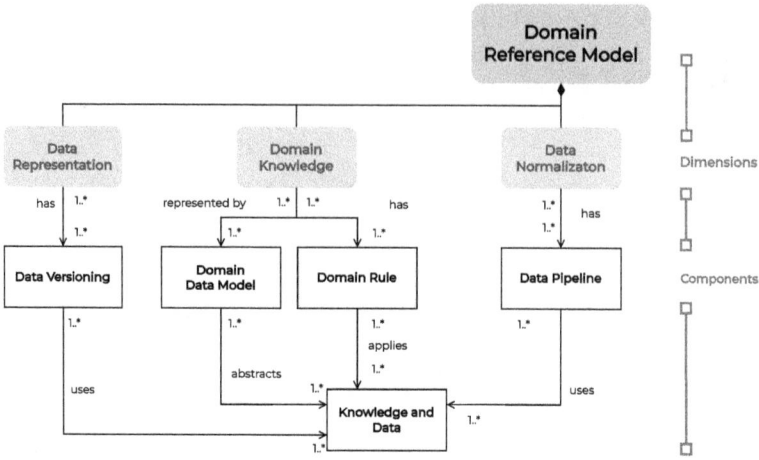

Fig. 1. Proposed model for inferring DRMs

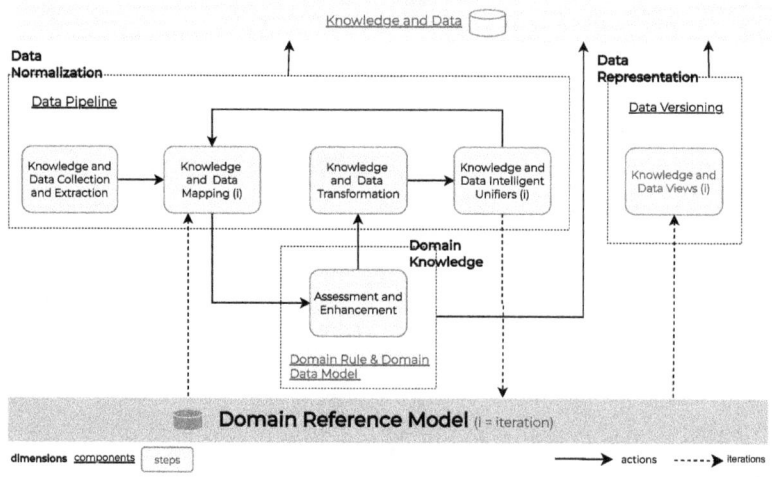

Fig. 2. Proposed method towards evolving DRMs

models; (2) Ensure the DRM dynamically adapts to new knowledge and domain requirements through iterative refinement.

To instantiate and validate the proposed method, each component and its associated steps will be instantiated using established concepts and tools, as depicted in Fig. 3. The steps are described as follows:

The first step collects domain-specific data and knowledge. This step focuses on gathering data from diverse sources, such as CSV files, JSON structures, and textual documents. Advanced techniques may include LLMs and Natural Language Processing (NLP), are employed to extract valuable information. User-defined functions and custom scripts further enhance the extraction step, producing outputs that include entities, rules, relationships, and properties.

Once the data is collected and extracted, the method moves to the Knowledge and Data Mapping step. Here, inputs such as vocabularies, rules, schema, expert knowledge and the current version of the DRM are processed to create mappings between the extracted data and predefined data models. This step may rely on third-party libraries and custom functions to align the data with domain-specific standards (such as NGSI-LD), ensuring consistency and normalization. The goal of this step is to create mappings that ensure the input data is aligned with a specific standard, enabling normalization for further transformations and integrations. The mappings are then assessed and enhanced (Domain Knowledge dimension) to ensure their correctness and relevance.

In the Knowledge and Data Transformation step, the mapped data is transformed into a more structured and usable format. This transformation involves converting the data into a KG, which organizes entities and their relationships in a way that supports further analysis and integration. The use of third-party libraries and advanced AI-driven transformation functions, support the data labeling and structuring, making it easier to align with the current DRM.

The final step of the Data Pipeline component is Knowledge and Data Intelligent Unifiers, it merges the previously transformed model with the DRM, addressing integration challenges, (such as alignment and completion). The result is assessed to ensure its validity supported by the Knowledge Dimension. The domain experts have an important role in this step, reviewing the outputs to identify and address any inconsistencies, outdated information, or newly created knowledge. The goal is to maintain a consistent and reliable representation of the domain within the DRM.

The Data Representation dimension involves steps that should provide partial or full views of the DRM, depending on the specific needs of the business or analytical tasks. These views are created using third-party libraries and custom functions, providing tailored representations of the data. Partial views might focus on specific subsets of the data (e.g., sales trends or customer segments), while full views offer a comprehensive overview of the entire domain. This step is iterative, allowing for continuous refinement and adaptation as requirements evolve. At the core of the method is the data model of the DRM. The DRM is represented in a KG and is continuously updated through iterative cycles (denoted by "i"). It encapsulates the domain's knowledge, rules, and relationships, ensuring that all data and transformations align with the domain's intrinsic properties. The DRM is not static; it evolves as new data is integrated and as domain requirements change.

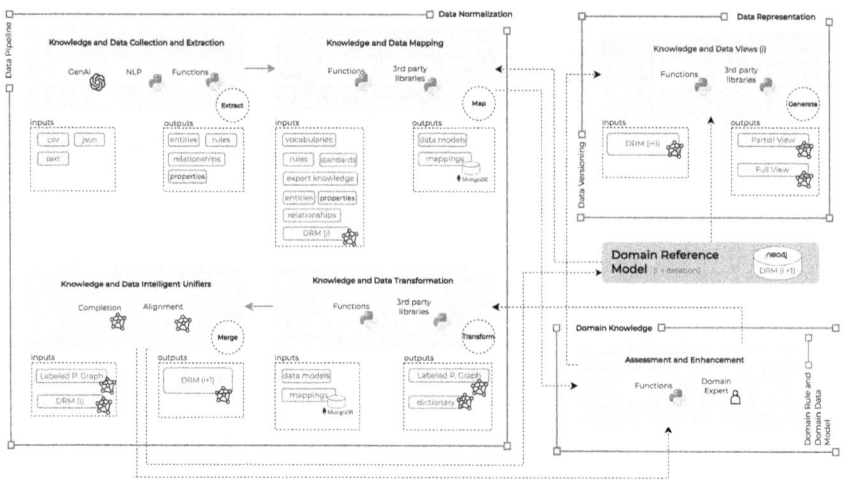

Fig. 3. Technological instantiation of the method towards evolving DRMs

6 Conclusion

This work proposes a structured model and method to systematically infer and evolve DRMs, ensuring alignment with domain knowledge while integrating diverse data sources. The proposed model identifies dimensions and components that must be considered in the inference of DRMs. By proposing a series of steps, the method ensures consistent alignment with domain and recognized data models and formats. Then, through iterative assessment and versioning, the model accommodates domain evolution, dealing with outdated or redundant data and knowledge. The next steps will focus on improving the literature review on DRMs to help refine the proposed model, and support the decomposition of dimensions, components, and steps. This will involve engaging domain experts through interviews to validate both the proposed model and method.

Acknowledgments. This research was supported by FCT - *Fundação para a Ciência e Tecnologia* within the R&D Units Project Scope: UIDB/00319/2020. I wish to express my sincere gratitude to Professor Maribel Y. Santos and Professor António C. Vieira for their invaluable supervision.

References

1. Burlton, R.: Delivering Business Strategy Through Process Management, pp. 5–37. Springer, Berlin (2010)
2. Fettke, P., Loos, P.: Classification of reference models - a methodology and its application. Inf. Syst. E-Bus. Manage. **1**, 35–53 (2003)
3. Gray, J., Rumpe, B.: Reference models: how can we leverage them? Softw. Syst. Model. **20**(6), 1775–1776 (2021)

4. Hevner, A.R., March, S.T., Park, J., Ram, S.: Design science in information systems research. MIS Q. **28**(1), 75–105 (2004)
5. Hofer, M., Obraczka, D., Saeedi, A., Köpcke, H., Rahm, E.: Construction of knowledge graphs: current state and challenges. Information **15**(8) (2024). https://doi.org/10.3390/info15080509
6. Hogan, A., et al.: Knowledge graphs. ACM Comput. Surv. **54**(4) (2021)
7. Ilyas, I.F., Rekatsinas, T., Konda, V., Pound, J., Qi, X., Soliman, M.: Saga: a platform for continuous construction and serving of knowledge at scale. In: Proceedings of the 2022 International Conference on Management of Data, SIGMOD 2022, pp. 2259–2272. Association for Computing Machinery, New York (2022)
8. Liu, J., Zhang, Q., Fu, L., Wang, X., Lu, S.: Evolving knowledge graphs. In: IEEE INFOCOM 2019 - IEEE Conference on Computer Communications. pp. 2260–2268 (2019)
9. Noy, N., Gao, Y., Jain, A., Narayanan, A., Patterson, A., Taylor, J.: Industry-scale knowledge graphs: lessons and challenges. Commun. ACM **62**(8), 36–43 (2019)
10. OASIS: Reference model for service oriented architecture 1.0 (2006)
11. Peffers, K., Tuunanen, T., Rothenberger, M., Chatterjee, S.: A design science research methodology for information systems research. J. Manage. Inf. Syst. **24**, 45–77 (2007)
12. Reinhartz-Berger, I., Soffer, P., Sturm, A.: Extending the adaptability of reference models. Trans. Sys. Man Cyber. Part A **40**(5), 1045–1056 (2010)
13. Rosemann, M.: Preparation of Process Modeling, pp. 41–78. Springer, Berlin (2003)
14. Santos, M.Y., Costa, C., Galvão, J., Andrade, C., Pastor, O., Marcén, A.C.: Enhancing big data warehousing for efficient, integrated and advanced analytics. In: Cappiello, C., Ruiz, M. (eds.) Information Systems Engineering in Responsible Information Systems, pp. 215–226. Springer International Publishing, Cham (2019)
15. Sjarov, M., Franke, J.: Towards knowledge graphs for industrial end-to-end data integration: technologies, architectures and potentials. Lect. Notes Prod. Eng. Part **F1160**, 545–553 (2022)
16. Tiddi, I., Schlobach, S.: Knowledge graphs as tools for explainable machine learning: a survey. Artif. Intell. **302** (2022)
17. Tu, H., Yu, S., Saikrishna, V., Xia, F., Verspoor, K.: Deep outdated fact detection in knowledge graphs. In: 2023 IEEE International Conference on Data Mining Workshops (ICDMW), pp. 1443–1452 (2023)
18. Wang, Q., Mao, Z., Wang, B., Guo, L.: Knowledge graph embedding: a survey of approaches and applications. IEEE Trans. Knowl. Data Eng. **29**(12), 2724–2743 (2017)
19. Yan, J., Wang, C., Cheng, W., Gao, M., Zhou, A.: A retrospective of knowledge graphs. Front. Comp. Sci. **12**(1), 55–74 (2018). https://doi.org/10.1007/s11704-016-5228-9

Tutorials

Problematising and Ideating for Design Science Research

John R. Venable(✉)

Curtin University, Bentley, Australia
j.venable@curtin.edu.au

Abstract. Problematisation is the activity of developing understanding, defining, and selecting the scope of a problem to be solved or improved upon by a Design Science Research (DSR) project. Ideation is the act of coming up with and selecting a new means for improving upon a selected and defined problem that further sets a DSR project's scope and will be developed and evaluated in the DSR project. Performing both activities well is essential for successful completion of a DSR project. This tutorial will define and frame these two activities within existing DSR processes and methodologies and develop practical skills for how to accomplish these activities. The principle method to be applied is Coloured Cognitive Map-ping for Design Science Research (CCM4DSR). Using CCM4DSR, DSR re-searches can analyse the causes and consequences of a problematic situation (using coloured cognitive maps or CCMs), transform the resulting understanding/CCM into a CCM representation of what it would mean to solve or improve upon the problem, ideate means to achieve the desired outcomes, decide upon the scope of a DSR project, select the attributes or criteria to evaluate, and select constructs for design knowledge theorising. CCM4DSR support for the last two activities (evaluation and theorising) will only be introduced, not developed further during the tutorial. Practical exercises will require attendees to use and discuss the frameworks and tools in CCM4DSR to develop skills in their use.

Taking Control of Our Privacy When Using Mobile Apps: Checks and Tools

M. Mercedes Martínez-González[1](✉), Alejandro Pérez-Fuente[1], Amador Aparicio[1], and Pablo-Abel Criado-Lozano[1,2]

[1] Privacy Engineering Research Group, Universidad de Valladolid, Valladolid, Spain
{mercedes,amador}@infor.uva.es, alejandro.perez.fuente@uva.es, pabloabel.criado24@estudiantes.uva.es
[2] Universidad Europea Miguel de Cervantes, Valladolid, Spain
https://ingpriv.uva.es/

Abstract. Mobile applications that provide access to information system services can impact user privacy. This impact influences the quality of an information system from perspectives such as ethics, regulatory compliance, and user satisfaction, ultimately determining their intention to use the system. In this tutorial, we address this issue from the user's perspective, working on users' ability to take action. From the understanding of the risks assumed when using mobile applications, to the use of indicators helpful to make informed decisions for self-protection.

Keywords: Privacy · Mobile Computing · Risk Assessment

Summary

Mobile applications that provide access to information system services have become essential for user interaction with these systems. However, we are not used to considering the impact of these applications on user privacy when assessing the risks associated with an information system. This is changing as this impact influences the quality of an information system from perspectives such as ethics, regulatory compliance, and user satisfaction, ultimately determining their intention to use the system.

In this tutorial, we will address this issue from the user's perspective. We will begin with a brief introduction to the issue and explore the extent of users' ability to take action. Next, we will demonstrate how to find and use quality indicators and information to make informed decisions. This helps users understand the risks they assume when using these applications and protect themselves effectively. Although the workshop is primarily focused on end users, the tools presented will also be useful to developers and other experienced users.

Expected Background of the Attendees

Attendees are expected to have a basic understanding of digital technologies, such as using mobile applications, navigating online services, and recognizing common security concepts. However, no prior expertise in cybersecurity, legal compliance, or software development is required. This tutorial is particularly valuable for end users seeking to enhance their privacy awareness, as well as for developers and IT professionals interested in better understanding user-centric privacy concerns and solutions.

Learning Objectives

By the end of the tutorial, participants will be able to:

LO1 Recognize the privacy risks users take when using mobile apps
LO2 Use the mechanism users have to protect their privacy in mobile apps
LO3 Use privacy indicators and quality repositories to assist decision-making
LO4 Assess the effect of their decisions as users on the privacy risks they take
LO5 Understand the potential of privacy metadata repositories for developers

The activities will be organized in four main phases. The relation of each phase with Learning Objectives and timeline is shown in Table 1:

Table 1. Structure: Contents, Learning Objectives, Timeline.

Content	Learning Objectives	Timeline
Introduction	LO1	20 min: Min. 1–20
Self Protection	LO2	20 min: Min. 21–40
Repositories & Indicators	LO3, L04, LO5	35 min: Min. 41–75
Conclusions	LO4, LO5	15 min: Min. 76–90

Methodology

An **active learning** approach will be employed, based on **case-based learning**. A set of real-world mobile applications will be selected as practical examples to work on throughout the tutorial.

Useful Links

ToS;DR	:	https://tosdr.org/
Exodus Privacy	:	https://exodus-privacy.eu.org/
APK Falcon (App-PI)	:	https://apkfalcon.inf.uva.es/

Acknowledgments. This tutorial is part of the activities of the research cybersecurity project "App-PI (App Privacy Impact)", a collaboration between the University of Valladolid and the INCIBE institute (Spanish Institute of Cybersecurity), supported by the Spanish "Plan de Recuperación, Transformación y Resiliencia", funded by the European Union's Next Generation programme.

Disclosure of Interests. The authors have no competing interests to declare that are relevant to the content of this tutorial.

Introduction to Fractal Enterprise Model (FEM) and FEM Toolkit

Ilia Bider[1,2]

[1] Stockholm University, Stockholm, Sweden
[2] University of Tartu, Tartu, Estonia
ilia@dsv.su.se

There exist several enterprise modeling languages, each aimed at a different usage. For example, ArchiMate can be used to understand what IT support exists or what IT supportis needed for certain business activities. Fractal Enterprise Model (FEM) can be used to understand how the enterprise functions as a whole in its environment, find missing activities, plan organizational changes, or radical business model transformations. FEM expresses the relationships between the enterprise assets, its business processes, and the enterpriseâĂŹs environment. The tutorial is aimed to introduce the participant to the FEMâĂŹs concepts: process and asset for describing enterprise activities, and external pool and external agent (e.g., a competitor) for describing the environment in which the enterprise operates. The tutorial will also introduce tools for drawing FEM diagrams (the FEM toolkit, which is freely available for download on Windows, Linux, and Mac) and presenting them on the WEB for stakeholders (the FEM viewer).

An example of an FEM diagram is shown in Fig. 1. It shows a model of a business consulting company. In the root of the diagram, there is a primary process (oval) - which represents the behavior. The shape has a double line border, which indicates that the primary process, i.e., the process has an external beneficiary who can pay for the service. Other ovals show supporting processes that are needed to provide the service. The processes are connected to the assets (set of things) - rectangle shapes - that are needed for running the processes repeatedly; the connections have the form of arrows with solid lines. The connection between the supporting processes and assets has the form of dashed arrows. A dashed arrow shows that the process is aimed to have the asset in working order. Arrows have labels that are predefined. A label on the solid arrow explains in what capacity the asset is used in a process. A label on the dashed arrow explains how the process manages the asset; there are three labels of this sort: *Acquire* - adding new elements; *Maintain* - maintaining elements in working order; and *Retire* - removing elements that can no longer be used in the process.

Two other shapes are meant to show the environment in which an enterprise operates. A cloud shape represents an external pool - a set of things - from which the enterprise can get its assets or to which it can add something, e.g., waste. The connection is shown by dashed blue arrows with a rounded tale. A rectangular shape with rounded corners represents an external actor that can be connected to the same pools. An external actor may represent competitors or collaborators.

There is a special Internet site [1] devoted to the Fractal Enterprise Model. From this site, it is possible to download the FEM toolkit [2] that will be used by the instructor to draw FEM diagrams. The toolkit exists for Windows, Linux, and Mac. The installation on Windows is straightforward. Mac installation can be a bit complicated. Write to the instructor in advance if you tried but were not lucky in installing the toolkit. There is also a web-based viewer [3] that allows to browse through the package of models, which will also be used in the tutorial.

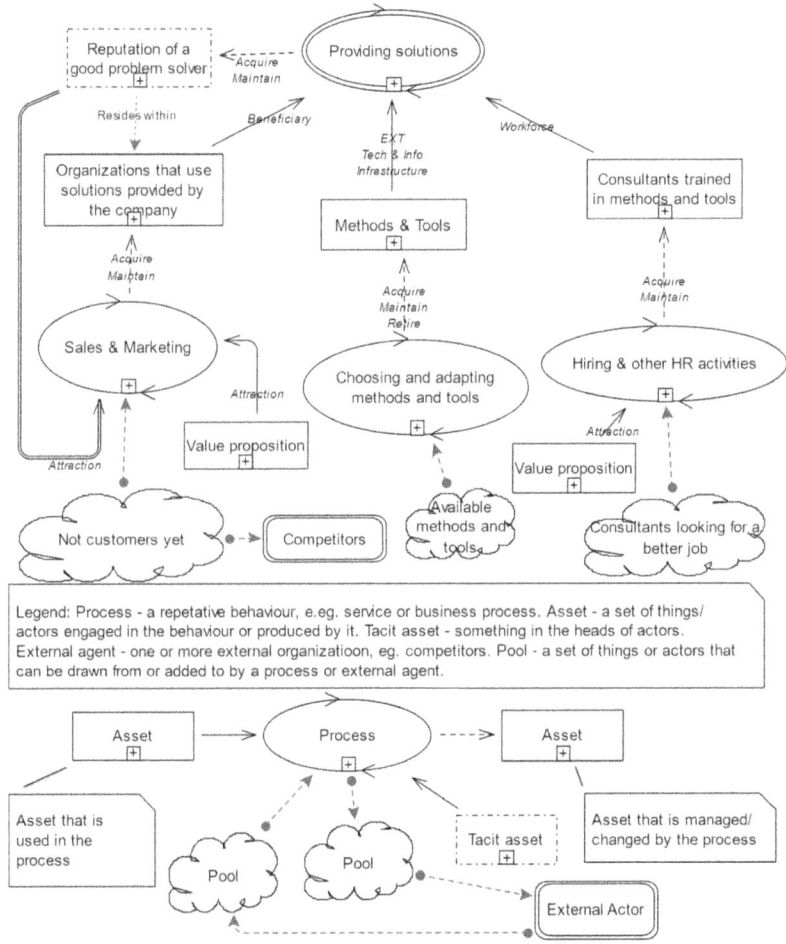

Fig. 1. An example of FEM model with some explanations

References

1. Fractalmodel.org: Fractal Enterprise Model. https://www.fractalmodel.org/. Accessed Mar 2025

2. Fractalmodel.org: FEM toolkit. https://www.fractalmodel.org/fem-toolkit/. Accessed Mar 2025
3. Fractalmodel.org: FEM viewer. https://www.fractalmodel.org/fem-viewer/. Accessed Mar 2025

Author Index

A
Abras, Jordan II-3
Aparicio, Amador II-229
Adamo, Greta I-156
Alfonso, Ivan II-130
Ali, Raian I-331, I-347
Almeida, João Paulo A. I-121
Alonso-Rocha, J. L. II-98
Alshakhsi, Sameha I-347
Al-Thani, Dena I-347
Aly, Adel I-262
Amous, Ikram I-37
Arden-Close, Emily I-331
Asensio, Estefanía Serral I-20

B
Babiker, Areej I-347
Babouchkine, Jean-Marc II-52
Balderas-Díaz, Sara I-52
Bangui, Hind I-87
Baños-González, Miguel II-151
Bártík, Jáchym I-245
Beerepoot, Iris I-71
Bergmann, Ralph I-20
Bertrand, Yannis I-20
Bider, Ilia II-120
Bonino da Silva Santos, Luiz Olavo I-103
Braakman, Mari A. J. I-71
Buchmann, Robert Andrei I-140
Buhnova, Barbora I-87
Burnay, Corentin II-3
Bider, Ilia II-231

C
Cabot, Jordi II-130
Calero, Coral II-151
Calhau, Rodrigo Fernandes I-121
Cancho-Casado, Jorge II-151
Carrión, Emilio I-279
Champagnat, Ronan I-3
Chen, Qian II-140
Cornax, Mario Cortes II-183
Corral-García, Javier II-151
Criado-Lozano, Pablo-Abel II-229

D
Dahache, Amira Ania I-3
Daoudi, Nadia II-130
Delgado-Pérez, Pedro I-52
Di Francescomarino, Chiara I-365
Di Mascolo, Maria II-52
Dias, Mariana II-20
Díaz, Oscar I-192
Dissegna, Sebastiano I-365
dos Reis, Gustavo Rodrigues II-183

E
El-Hajjami, Abdelkarim I-208
Enríquez, J. G. II-98

F
Fast Lappalainen, Katarina I-175
Faulkner, Stéphane II-3

G
García, Félix O. II-151
García-Romero, Manuel I-470
Garmendia, Xabier I-192
Ge, Mouzhi I-87
Gharbi, Mohamed II-52
González Enríquez, José I-470
Granda, Maria Fernanda I-313
Grüger, Joscha I-20
Gruninger, Michael II-109
Guerra, Eduardo I-103
Guerrero-Contreras, Gabriel I-52
Guimarães, Pedro II-218
Guizzardi, Giancarlo I-103, I-121
Guizzardi, Renata I-103
Gurgun, Selin I-331

H
Haug, Kerstin I-418
Henkel, Martin I-175
Hoffmann, Raquel I-121
Holubová, Irena I-245
Hübner, Karin I-226

J
Jiménez-Ramírez, A. II-98
Jiménez-Ramírez, Andrés I-470
Jmal, Raouf I-37
Joudieh, Noura I-3

K
Kampourakis, Konstantinos E. II-197
Kern, Christopher Julian I-226
Khraiwesh, Samira I-452
Kired, Nour Elhouda II-67
Knies, Eva I-71
Koupil, Pavel I-245
Krauze, Beāte II-162
Kroenung, Julia I-226
Küsters, Aaron I-383

L
Labbé, Cyril II-183
Lee, Suhwan I-401
Lindeberg, Jöran I-175
Logan, James I-121
Lopes Cardoso, Henrique II-89
Lopes, Carla Teixeira II-20
Lu, Xixi I-401

M
Malburg, Lukas I-20
Maldonado, Ricardo I-313
Martínez-Rojas, A. II-98
Martínez-Rojas, Antonio I-470
Masmoudi, Mariam I-37
Medina-Bulo, Inmaculada I-52
Moher, Riley II-109
Moraga, Mł Ángeles II-151
Moreira, João Rebelo I-121
Mos, Adrian II-183
Mughaz, Dror II-38
Martínez-González, M.Mercedes II-229

N
Nuñez, Ana-Gabriela I-313

O
Oliveira, Ítalo I-103

P
Palacio, Ana León I-295
Parra, Otto I-313
Peeters, Maria I-71
Phalp, Keith I-331
Pires, Luís Ferreira I-121
Pivert, Olivier I-262
Poss, Leo I-226
Proper, Henderik I-103
Pufahl, Luise I-452
Pulido, Carlos II-151
Pérez-Fuente, Alejandro II-229

R
Ravat, Franck II-67
Reijers, Hajo A. I-71, I-401
Rinderle-Ma, Stefanie I-418, II-140
Rodrigues, José Frederico II-89
Ronzani, Massimiliano I-365

S
Saarsen, Toomas II-120
Sales, Tiago Prince I-103
Salinesi, Camille I-208
Sarmiento, María-Belén I-313
Schnellmann, Marianne II-207
Schönig, Stefan I-226
Schultheis, Alexander I-20
Schumacher, Pol I-418
Sèdes, Florence I-37
Sidorova, Natalia I-435
Song, Jiefu II-67
Sperotto, Anna I-156
Suire, Cyrille I-3
Supti, Tourjana Islam I-347
Susaiyah, Allmin I-435

T
Teixeira Lopes, Carla II-89
Teste, Olivier II-67

Author Index

Thion, Virginie I-262
Trabelsi, Marwa I-3

U
Uifălean, Ștefan I-140

V
Valderas, Pedro I-279
Valle, Isadora I-103
van Wingerden, A. B. II-172
van der Aalst, Wil M. P. I-383

Verdier, Christine II-52
Venable, John R. II-228

W
Weigand, H. II-172
Wen, Lijie II-140
Willis, Max I-156
Wittges, Holger I-418

Y
Yankouskaya, Ala I-347